CRUISE MISSILES AND U.S. POLICY

Studies in Defense Policy
TITLES IN PRINT

Support Costs in the Defense Budget: The Submerged One-Third
Martin Binkin

The Changing Soviet Navy
Barry M. Blechman

Strategic Forces: Issues for the Mid-Seventies
Alton H. Quanbeck and Barry M. Blechman

U.S. Reserve Forces: The Problem of the Weekend Warrior
Martin Binkin

U.S. Force Structure in NATO: An Alternative
Richard D. Lawrence and Jeffrey Record

U.S. Tactical Air Power: Missions, Forces, and Costs
William D. White

U.S. Nuclear Weapons in Europe: Issues and Alternatives
Jeffrey Record with the assistance of Thomas I. Anderson

The Control of Naval Armaments: Prospects and Possibilities
Barry M. Blechman

Stresses in U.S.-Japanese Security Relations
Fred Greene

The Military Pay Muddle
Martin Binkin

Sizing Up the Soviet Army
Jeffrey Record

Modernizing the Strategic Bomber Force: Why and How
Alton H. Quanbeck and Archie L. Wood with the assistance of Louisa Thoron

Where Does the Marine Corps Go from Here?
Martin Binkin and Jeffrey Record

Deterrence and Defense in Korea: The Role of U.S. Forces
Ralph N. Clough

Women and the Military
Martin Binkin and Shirley J. Bach

The Soviet Military Buildup and U.S. Defense Spending
Barry M. Blechman and others

Soviet Air Power in Transition
Robert P. Berman

Shaping the Defense Civilian Work Force: Economics, Politics, and National Security
Martin Binkin with Herschel Kanter and Rolf H. Clark

The Military Equation in Northeast Asia
Stuart E. Johnson with Joseph A. Yager

Youth or Experience? Manning the Modern Military
Martin Binkin and Irene Kyriakopoulos

Paying the Modern Military
Martin Binkin and Irene Kyriakopoulos

Defense in the 1980s
William W. Kaufmann

The FX Decision: "Another Crucial Moment" in U.S.-China-Taiwan Relations
A. Doak Barnett

Planning Conventional Forces: 1950–80
William W. Kaufmann

Blacks and the Military
Martin Binkin and Mark J. Eitelberg with Alvin J. Schexnider and Marvin M. Smith

U.S. Arms Sales: The China-Taiwan Tangle
A. Doak Barnett

Cruise Missiles and U.S. Policy
Richard K. Betts

STUDIES IN DEFENSE POLICY

CRUISE MISSILES AND U.S. POLICY

Richard K. Betts

THE BROOKINGS INSTITUTION
Washington, D.C.

1775 Massachusetts Avenue, N.W., Washington, D.C. 20036

Library of Congress Catalog Card Number 82-072704
ISBN 0-8157-0933-1

9 8 7 6 5 4 3 2 1

THE BROOKINGS INSTITUTION is an independent organization devoted to nonpartisan research, education, and publication in economics, government, foreign policy, and the social sciences generally. Its principal purposes are to aid in the development of sound public policies and to promote public understanding of issues of national importance.

The Institution was founded on December 8, 1927, to merge the activities of the Institute for Government Research, founded in 1916, the Institute of Economics, founded in 1922, and the Robert Brookings Graduate School of Economics and Government, founded in 1924.

The Board of Trustees is responsible for the general administration of the Institution, while the immediate direction of the policies, program, and staff is vested in the President, assisted by an advisory committee of the officers and staff. The by-laws of the Institution state: "It is the function of the Trustees to make possible the conduct of scientific research, and publication, under the most favorable conditions, and to safeguard the independence of the research staff in the pursuit of their studies and in the publication of the results of such studies. It is not a part of their function to determine, control, or influence the conduct of particular investigations or the conclusions reached."

The President bears final responsibility for the decision to publish a manuscript as a Brookings book. In reaching his judgment on the competence, accuracy, and objectivity of each study, the President is advised by the director of the appropriate research program and weighs the views of a panel of expert outside readers who report to him in confidence on the quality of the work. Publication of a work signifies that it is deemed a competent treatment worthy of public consideration but does not imply endorsement of conclusions or recommendations.

The Institution maintains its position of neutrality on issues of public policy in order to safeguard the intellectual freedom of the staff. Hence interpretations or conclusions in Brookings publications should be understood to be solely those of the authors and should not be attributed to the Institution, to its trustees, officers, or other staff members, or to the organizations that support its research.

FOREWORD

ONE OF President Reagan's first initiatives was to increase the defense budget sharply. Fiscal constraints, however, raise questions about how much of the planned military buildup can be sustained. At the same time, a whole new family of weapon systems—cruise missiles—is reaching maturity. These missiles are potentially applicable to a wide range of military missions—strategic nuclear war, theater nuclear operations, attacks on land targets with conventional munitions, and naval engagements.

Although cruise missiles offer new ways of increasing U.S. military power, uncertainty about the resources that will be available in the future necessitates careful choices about which programs should have priority. In the missile's development phase and recent administration planning, enthusiasm raced ahead of analysis that would define the most promising versions of the new weapon and the versions for which other types of weapons, such as tactical aircraft, torpedoes, or rocket-powered missiles, are effective substitutes. Many observers see in cruise missiles an inexpensive solution to numerous military problems, yet they will absorb a significant part of the defense budget. More analysis of the missiles' comparative effectiveness for different missions is needed if the budget is to be allocated wisely.

In the interest of clarifying the background and significance of cruise missile programs, the interacting technological, strategic, and political factors that have shaped their development, and the critical trade-offs they pose for defense decisionmaking, Brookings published *Cruise Missiles: Technology, Strategy, Politics,* a 612-page book edited by Richard K. Betts, in 1981. This staff paper is a brief summary of that book, updated wherever possible.

Richard K. Betts is a senior fellow in the Brookings Foreign Policy

Studies program. The institution is grateful to the Ford Foundation for financial support that made this and the previous study possible.

The views expressed in this paper are those of the author and should not be ascribed to the Ford Foundation or to the trustees, officers, or other staff members of the Brookings Institution.

BRUCE K. MACLAURY
President

August 1982
Washington, D.C.

Abbreviations and Acronyms

ABM	antiballistic missile
ALCM	air-launched cruise missile
CEP	circular error probable
DSMAC	digital scene matching area correlator
GLCM	ground-launched cruise missile
ICBM	intercontinental ballistic missile
LRTNF	long-range theater nuclear forces
MIRV	multiple independently targetable reentry vehicle
MRASM	medium-range air-to-surface missile
SALT	strategic arms limitation talks
SIOP	single integrated operations plan
SLBM	submarine-launched ballistic missile
SLCM	sea-launched cruise missile
SRAM	short-range attack missile
SSN	nuclear attack submarine
START	strategic arms reduction talks
TASM	Tomahawk antiship missile
TEL	transporter-erector-launcher
TERCOM	terrain contour matching
TLAM-C	conventionally armed Tomahawk land-attack missile
TLAM-N	nuclear-armed Tomahawk land-attack missile

CRUISE MISSILES AND U.S. POLICY

Richard K. Betts

AFTER more than a decade of development the long-range cruise missile is about to burst upon the military scene. Recent administration plans call for the procurement of almost 9,000 cruise missiles of various models over the coming years, at a combined total cost of about $24 billion.[1] The cruise missile is a small and comparatively uncomplicated weapon system, but its consequences are likely to be large and complex. The missile arrives at a crucial juncture in foreign and defense policy and clearly will have major effects on U.S. military forces, prospects for arms control negotiations with the Soviet Union, and the cohesion of the Western alliance. Exactly what those major effects will be, however, is far less clear. This combination of importance and uncertainty is particularly troubling because the direction of U.S. defense policy is once again being debated heatedly; a faltering economy focuses attention on rethinking the size of budget increases and the public is extremely uneasy about nuclear strategy. As the cruise missile touches so many areas of policy, it will also figure in proposals for paring some defense initiatives or revising strategies.

The need for greater clarity in assessing the future role of the cruise missile emerges from several problems.

First, the genesis of the weapon was rather disorderly. Originally none of the military services, with the partial exception of the Navy, wanted it. Development of the missile was promoted by civilian leaders

1. The Defense Department's "Selected Acquisition Reports as of December 31, 1981," estimated the combined costs of the sea-, air-, and ground-launched cruise missile programs at over $23.9 billion. This total does not include potential expenditures for another cruise missile program—the medium-range air-to-surface missile.

for various reasons, only some of which were grounded in operational concerns. Secretary of State Henry Kissinger, for example, wanted it as a bargaining chip to barter away in arms control negotiations with Moscow. The progress of the weapon's development, however, turned it into a far more significant set of programs than most officials had dreamed of at its inception. The cruise missile is now the technological innovation that will most affect offensive military firepower in the 1980s. Other new weapons are evolutionary improvements of existing types, such as ballistic missiles or manned aircraft. Yet this major consequence is the product almost as much of inadvertence as of design.

Second, the cruise missile is an extraordinarily versatile system, applicable to a surprisingly wide range of missions. It can be deployed in various forms and launched from almost anywhere—from vehicles on the ground, from planes in the air, from surface ships, or from submarines. Because it is so adaptable, inventive planners have been devising ways in which it can be used for a multiplicity of military purposes. These ideas, however, have outpaced analysis of whether other systems might more effectively be used for some of these purposes.

Third, because the missile can carry either nuclear or conventional munitions,[2] it is not only tactically versatile but strategically significant as well; its potential use spans the entire range of U.S. strategy—strategic deterrence, theater nuclear war, conventional attacks on land targets, and naval engagements. For a defense planner this potential offers both great opportunity and great difficulty, since maximal exploitation is bound to complicate possibilities for arms control agreements—agreements that, if sensibly designed, could enhance security by limiting Soviet capabilities.

Fourth, the cruise missile has become a sensitive political issue not only with the USSR but with American allies as well. At the end of 1979 it was offered to allay Western European governments' anxiety about the shifting military balance. Since then, however, it has become a focus of domestic antinuclear opposition in several NATO countries. A military solution has become a political hot potato.

In short, the technological possibilities offered by the cruise missile spurred major commitments to procurement before the rationales for some of the programs were completely clear, and without a publicly articulated, comprehensive assessment of costs and benefits. Before all

2. According to one report, consideration has even been given to putting chemical warheads on ground-launched cruise missiles. Walter Pincus, "Cruise Missiles May Get Chemical Warheads," *Washington Post*, January 16, 1982.

final decisions are made as programs evolve in the next several years, and to avoid potential pitfalls from decisions already made, clearer assessment is necessary.

This staff paper surveys the complexities and interrelationships of the technological, strategic, and political implications of cruise missile development. It does not incorporate the full analysis supporting its conclusions but is drawn for the most part (although it includes supplementary and updated material) from a much larger study, to which interested readers can refer for elaboration of particular issues.[3]

Technology

Modern long-range cruise missiles are distinctively new weapons, although crude versions of the basic concept first appeared almost forty years ago.[4] Before the future consequences of current programs are discussed, it is important to define how the modern cruise missile evolved, what its unique capabilities are, and what makes it more attractive than other weapon systems that could be used for the same tasks.

Definition and Background

A cruise missile is an unmanned, air-breathing, expendable vehicle programmed to deliver an explosive charge against a target. Because the propulsion system (like that of jet planes) uses air as an oxidizer, the missile can fly further than a rocket-powered system carrying the same amount of fuel. Because it is expendable (meaning it flies only one mission, crashing into its target), a cruise missile can be built with less expensive materials than a strike aircraft (which must hold up over years of use) and need not incorporate all of the complex electronic or defensive systems of a manned aircraft.[5]

3. Richard K. Betts, ed., *Cruise Missiles: Technology, Strategy, Politics* (Brookings Institution, 1981), 612 pages. This staff paper is drawn primarily from my Introduction and Conclusion to the above volume but includes some summary material from chapters by the other authors.

4. This study is not concerned with shorter-range cruise missiles, such as the Navy's Harpoon, which has been in service for some time. Nor is it concerned with older, less sophisticated cruise missiles, such as those deployed for several decades by the Soviet Navy.

5. For full discussion of the missile's workings, see John C. Toomay, "Technical Characteristics," in Betts, ed., *Cruise Missiles*, pp. 31–52.

The simple cruise missile (Germany's V-1 buzz bomb) made its debut in World War II. At first it was quite effective, but because the technology was crude—the V-1 was large and slow and flew at relatively high altitudes—English air defenses were able to adapt and eventually destroyed the bulk of the missiles before they could reach their targets.[6] After World War II the Soviet Union developed naval cruise missiles (largely as a counter to U.S. aircraft carriers).[7] In the 1950s the United States deployed some crudely guided cruise missiles for strategic forces, but they never became reliable enough to compete with emerging ballistic missile systems. The Snark cruise missile, for example, was a fiasco: "The average miss distance was over 1,000 miles. At least one came down in the wrong hemisphere, disappearing somewhere in the interior of Brazil."[8]

What transformed the neglected cruise missile into an important part of defense programs was an uncoordinated evolution of technological innovations rather than a deliberate effort or an epochal breakthrough. Several developments that reached fruition around 1970 made the cruise missile concept more attractive: small efficient turbofan engines; small high-yield warheads; more accurate mapping data; and less radar-reflective airframes. These improvements allowed exploitation of more accurate guidance systems.

The terrain contour matching (TERCOM) guidance system was actually invented in the late 1950s. TERCOM uses a radar altimeter and computer to match readings of terrain gradients on the ground over which it flies with internally stored maps. Inertial guidance carries the missile to the land area of the first map, where TERCOM then corrects the missile's course. This process can be repeated numerous times, and it also allows a zigzag course of flight to evade defenses (see figure 1). TERCOM permits a circular error probable (CEP—a measure of accuracy) of close to 300 feet.

Because nonnuclear ordnance requires greater accuracy to destroy a target, guidance for conventionally armed cruise missiles is supplemented with the digital scene matching area correlator (DSMAC) system, which permits a CEP reportedly as low as a few feet. DSMAC is more

6. James Phinney Baxter 3d, *Scientists Against Time* (Little, Brown, 1950), pp. 234–35.

7. See Raymond L. Garthoff, "Soviet Perspectives," in Betts, ed., *Cruise Missiles*, pp. 339–42.

8. Edmund Beard, *Developing the ICBM: A Study in Bureaucratic Politics* (Columbia University Press, 1976), p. 224, n. 15.

Figure 1. Flight Path of Missile Using Terrain Contour Matching (TERCOM) Guidance

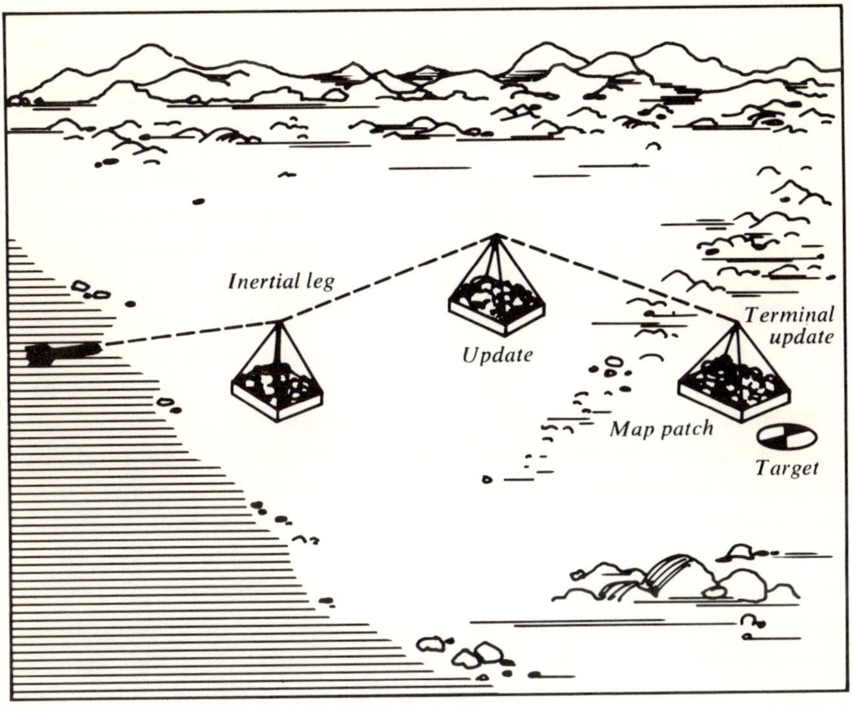

sensitive, relying on a digitized picture of the ground, and therefore has limitations at night, in bad weather, or after seasonal change.

Because its small size affords lower radar cross section and its ability to fly close to the ground allows it to "hide" behind hills and the curvature of the earth, the cruise missile is much harder for enemy radars to detect than large manned aircraft. At least until the Soviet Union perfects a look-down–shoot-down defense system and deploys it in large numbers (and to a lesser degree even afterward), U.S. cruise missiles have a high probability of penetrating to strategic targets in the Soviet homeland. Large area targets (such as cities) are difficult to protect when the missiles' flight paths are not known and the number of defense sites is limited, especially since defenses can be saturated with a large number of missiles. Even a low "leakage" rate of defenses ensures that nuclear-armed missiles will inflict substantial damage on area targets. Terminal defense of point targets, however, is much easier.[9] Moreover, the density

9. For detailed analysis, see Gordon MacDonald, Jack Ruina, and Mark Balaschak, "Soviet Strategic Air Defense," in Betts, ed., *Cruise Missiles*, pp. 53–82.

of Soviet *tactical* air defenses reduces the assurance of cruise missile penetration against some military targets in a potential war in central Europe.

Current Programs

Development of the modern cruise missile began in 1972. Originally focused on the option of a submarine-launched nuclear weapon, the program evolved into a much wider range of weapons and launch modes.[10] The major programs currently planned are noted below and their technical characteristics are summarized in table 1.

THE AIR-LAUNCHED CRUISE MISSILE (ALCM). Under the Reagan administration's plan, 4,348 Boeing AGM-86B ALCMs are to be procured and deployed in stages (beginning at the end of 1982) on B-52 bombers. (This represents over a 25 percent increase from the Carter administration's plan for 3,418 ALCMs.) Initially the B-52s will carry six ALCMs on each of two wing pylons, retaining gravity bombs and short-range attack missiles (SRAMs) in the internal bomb bay. In this phase the bombers will operate in a "shoot and penetrate" mode, launching ALCMs from beyond Soviet defenses, then flying on to use their other weapons. After late 1985 these bombers will be converted fully to cruise missile carriers with internal rotary launcher racks. The total cost of the ALCM program as estimated at the beginning of 1982 (including costs for modifying the B-52s) was approximately $8.1 billion.[11] Reagan administration plans for the procurement of B-1B penetrating bombers envision that they will also be converted to carry cruise missiles when newer Stealth bombers enter the force in the next decade.

THE GROUND-LAUNCHED CRUISE MISSILE (GLCM). Four hundred and sixty-four nuclear-armed GLCMs are due to be deployed in Europe beginning in December 1983 as part of NATO's 1979 decision to modernize long-range theater nuclear forces (LRTNF), although political opposition in Western Europe or potential arms control agreements could change these plans. The GLCMs are to be mobile, in groups of sixteen missiles—four on each transporter-erector-launcher (TEL)—in a convoy of vehicles with launch control centers and other support elements. Total program cost is estimated at about $3.3 billion.[12]

10. See Ron Huisken, "The History of Modern Cruise Missile Programs," in ibid., pp. 83–100.

11. "Selected Acquisition Reports as of December 31, 1981."

12. Ibid.

Table 1. U.S. Cruise Missile Programs

Missile[a]	*Range (nautical miles)*	*Ordnance*	*Carrier/ launcher*	*Number*[b]	*Cost*[c]
ALCM	1,500	Nuclear	B-52 bomber	4,348	8.1
GLCM	1,300–1,500	Nuclear	Truck convoy	560	3.3
SLCM					
TLAM-N	1,300–1,500	Nuclear	Submarine or surface ship	4,000	12.6
TLAM-C	700	Conventional			
TASM	300	Conventional			
MRASM	300	Conventional	Tactical aircraft	?	?

Sources: John C. Toomay, "Technical Characteristics," in Richard K. Betts, ed., *Cruise Missiles: Technology, Strategy, Politics* (Brookings Institution, 1981), p. 46; Department of Defense, "Selected Acquisition Reports as of December 31, 1981."

a. Abbreviations are defined in the glossary facing page 1.

b. Planned procurement.

c. Billions of dollars.

THE TOMAHAWK SEA-LAUNCHED CRUISE MISSILE (SLCM). The U.S. Navy plans to procure about 4,000 General Dynamics Tomahawk missiles for deployment in three modes, using 2,600 launchers aboard 150 naval combatants. No decision on the mix—how many of each version will be deployed to make up the total—has been announced. Total estimated program cost, including new vertical launch systems and supporting programs, is about $12.6 billion.[13] Three basic versions of the SLCM are to be deployed.

The Tomahawk land-attack missile, nuclear-armed (*TLAM-N*). Plans for the TLAM-N, one of several versions of the sea-launched cruise missile, are currently fluid and controversial. The status of this variant was uncertain at the end of the Carter administration, but President Reagan has announced that it will be deployed. Most will probably be fitted in new vertical launchers, though some will be fired from torpedo tubes.

The Tomahawk land-attack missile, conventionally armed (*TLAM-C*). This model can be fitted on surface ships and submarines, fired from torpedo tubes, armored box launchers, or a vertical launching system now in development. The TLAM-C may be deployed in 1982.

The Tomahawk antiship missile (*TASM*). This has longer range and larger payload (up to 1,000 pounds) than the Harpoon cruise missile,

13. Ibid.; "Quantities, Inflation Indices Boost Costs of Weapons," *Aviation Week and Space Technology*, April 12, 1982, p. 64; Lieutenant Paul G. Johnson, USN, "Tomahawk: The Implications of a Strategic/Tactical Mix," *U.S. Naval Institute Proceedings*, vol. 108 (April 1982), p. 28.

which is currently deployed. It is scheduled to be operational in mid-1982. Because TERCOM does not work over water or against mobile targets, TASMs are fitted with seekers that allow them to be launched over the horizon, fly a search pattern, distinguish friendly from enemy ships, and attack the latter with high accuracy.

THE MEDIUM-RANGE AIR-TO-SURFACE MISSILE (MRASM). For joint use by the Navy and the Air Force, this weapon is in development and could be launched from a variety of tactical aircraft against sea and land targets.

Plans for the production and deployment of the ALCM and the GLCM are the firmest at present. The scale and mode of deployment for the other models are still subject to uncertainties about development, testing, cost, strategy, arms control, European domestic politics, and possibilities that other classes of weapons, such as rocket-powered missiles or manned aircraft, may be preferable to cruise missiles for some missions.

Force Structure, Doctrine, and Policy

The programs described evolved because technology made them possible, not because the professional military believed they were necessary. The services for the most part resisted cruise missiles for fear they might mean reduced commitment to other, preferred weapon systems. Development was pushed by civilian technical specialists, systems analysts, and managers.[14] Although the civilian sponsors had strategic and tactical rationales in mind, the result is that major changes in force structure are being made before the role of the weapons in military operational doctrine has been fully articulated.

It is not necessarily irrational for changes in force structure to precede doctrinal adaptation. Strategists usually lack the scientific expertise to levy realistic requirements for innovative types of weapons, and scientific serendipity may provide options that strategists cannot imagine in advance. In any case, most new weapons are adopted well before doctrine is adapted. For example, radical changes in the role of machine guns in World War I and of tanks in World War II occurred long after those weapons had first been deployed. Adapting doctrine to technology, rather than the other way around, may be particularly sensible if the weapon system is very versatile and the doctrine puts a high premium on operational flexibility.

14. See Robert J. Art and Stephen E. Ockenden, "The Domestic Politics of Cruise Missile Development, 1970–1980," in Betts, ed., *Cruise Missiles*, pp. 359–93.

Two essential reasons for the appeal of cruise missiles are their flexibility, which makes them useful for many purposes, and their comparatively low cost, which makes it possible to deploy them in great quantity. In principle this is an ideal combination, especially since the defense community has become anxious about the tendency in American weapon development to sacrifice quantity to quality, leaving forces with progressively smaller numbers of sometimes excessively sophisticated systems.

While versatile, however, cruise missiles are not a military panacea, and their cost advantage is not clear in all cases; they are not necessarily valid competitors to other systems for all military missions. Some early expectations were excessive because operational constraints were misunderstood, and some ideas about the ultimate potential for variants of these systems bordered on the euphoric or even the fantastic. Two analysts, for example, argued that reducing the flight range of bombers carrying the missiles would reduce tanker requirements (as long as the missiles are carried on external pylons, however, greater air drag will *increase* tanker requirements); and looking ahead to an advanced conventional cruise missile, they rhapsodized that "if developed to employ multiple anti-armor warheads, as few as four cruise missiles could search out and destroy up to 200 enemy tanks."[15] Recent testing has led to more modest assumptions about cruise missile virtuosity, at least in the first generation.[16] Technological uncertainties, though, cut in both directions: variation, refinement, and combination of advances may produce remarkably new sorts of weapons.[17] The cruise missile became appealing despite the greater unit effectiveness of other systems (such as ballistic missiles and tactical aircraft) for two principal reasons: first, during the

15. Robert L. Pfaltzgraff, Jr., and Jacquelyn K. Davis, *The Cruise Missile: Bargaining Chip or Defense Bargain?* (Cambridge: Institute for Foreign Policy Analysis, 1977), pp. 21, 36. If all the strategic air-launched missiles were carried internally to eliminate the extra drag, the number of deliverable warheads per carrier would be reduced. For another optimistic view of the potential of conventional cruise missiles against mobile targets, see "Cruise Missiles with Laser Radars Studied," *Aviation Week and Space Technology*, December 1, 1980, p. 141.

16. See *Most Critical Testing Still Lies Ahead for Missiles in Theater Nuclear Modernization,* Report to the Congress by the Comptroller General of the United States, MASAD-81-15 (General Accounting Office, 1981), pp. 14–15.

17. A report on the U.S. Air Force study Strategic Missile System (SMS) 2000, which projects *ballistic* missile technology beyond the year 2000, suggests that "a hybrid cruise missile/ballistic system could loiter while operating in the air-breathing mode to assure survivability over protracted periods. Upon receipt of the 'go code,' the weapon transitions to ballistic missile operation to provide for a rapid and reliable penetration capability." Edgar Ulsamer, "In Focus . . ." *Air Force,* February 1981, p. 23.

crucial years of research and development (through the mid-1970s) it was not constrained by SALT agreements; second, and most important, it seemed inexpensive and thus susceptible to deployment in large numbers.[18]

Versatility, however, makes some cruise missile variants difficult to accommodate within arms control schemes, and arms control—if advantageously negotiated—can be one means (by constraining the adversary's forces) of stabilizing security at lower cost than would be necessary with unrestrained competition in deployments. Some weapon developments that solve military problems in the short term can worsen them in the long term: multiple independently targetable reentry vehicles (MIRVs) are an example.[19] The central problem is to judge the interlocking military and political opportunity costs of different cruise missile programs. On one hand, it is vital to be sure that deployments which might work against arms control gain more in U.S. offensive capability than they would lose by increased vulnerability to Soviet power should Moscow deploy more of its own offensive systems than it would if constrained by an arms limitation treaty. On the other hand, cruise missiles' potential military value must also be weighed against that of other weapons that could be procured with the same funds and that could perform similar missions. To understand which roles for cruise missiles are most promising and which could be dispensed with to allow investment in alternate weapons or bargaining for worthwhile Soviet concessions in arms control negotiations, more discussion of military applications is necessary.

18. "Because the weapon was a one-shot deal, expected to perform only once and then for a few hours at most, cheaper materials, lower manufacturing tolerances, and other shortcuts could be made in design and production. The absence of a crew member and associated safety devices, instrumentation and safety factors in engine and airframe further cut weight, complexity, and cost." Kenneth P. Werrell, "The Cruise Missile: Precursors and Problems," *Air University Review*, vol. 32 (January–February 1981), pp. 36–38.

19. While MIRVs were deployed by the United States alone they offered an inexpensive means of multiplying the number of U.S. warheads, thus countering the steady Soviet buildup in single-warhead intercontinental ballistic missiles. When the Soviet Union deployed MIRVs, however, the innovation worked against U.S. interests by increasing the vulnerability of U.S. missiles. Parity in single-warhead fixed ICBMs favors the side striking second (and is thus theoretically stabilizing) because a first strike would disarm the attacker more than the victim. MIRVs reverse the calculus, since an attacker can use a small number of its own launchers to destroy a large number of the victim's. Thus if an arms control agreement to ban MIRVs had been feasible, the ICBM component of the U.S. deterrent would have remained more secure than it now is. On the failure to pursue a MIRV ban, see Gerard Smith, *Doubletalk: The Story of the First Strategic Arms Limitation Talks* (Doubleday, 1980), pp. 154–78, 415–17, 432, 471–72.

Nuclear Strategy

Cruise missiles influence nuclear strategy in two related dimensions: the "balance" of forces as perceived in peacetime, and operational employment in the event of war. The former is central to stability and deterrence since it affects political decisionmakers' behavior. The latter is important because the credibility of war plans reinforces deterrence.

Strategic Forces: Goals and Capabilities

The principal official guidelines for the adequacy of strategic forces in recent years, in ascending order of the demands they impose, have been three.

Assured destruction: the capacity to retaliate and inflict unacceptable damage upon the Soviet Union's economy and population after a preemptive attack. As a *minimum* requirement, this is the least controversial criterion. Although its influence has plummeted in recent years, there is still a lobby, which contains some eminent spokesmen, that views assured destruction as close to a *sufficient* standard.[20] This view may gain in influence once again if the popular anxiety about nuclear weapons that has grown so notably in 1982 endures.

Essential equivalence: according to Secretary of Defense Harold Brown, "a condition such that any advantages in force characteristics enjoyed by the Soviets are offset by other U.S. advantages. Although we must avoid a resort to one-for-one matching of individual indices of capability, our strategic nuclear posture must not be, and must not seem to be, inferior in performance to the capabilities of the Soviet Union." This is a psychological and political as well as operational criterion and is measured primarily in terms of static indexes.[21]

Countervailing strategy: a flexible range of selective force employment options. These include an increased capacity to destroy military forces and command centers, to match or counter a number of conceivable Soviet attack plans, to deny the attainment of any strategic objective

20. See McGeorge Bundy, "Maintaining Stable Deterrence," *International Security,* vol. 3 (Winter 1978–79), pp. 6–10, 13.

21. *Department of Defense Annual Report, Fiscal Year 1979* (Government Printing Office, 1978), p. 56; *Department of Defense Annual Report, Fiscal Year 1981* (GPO, 1980), pp. 68–69.

such plans might envisage, and to retain substantial forces in reserve. In former Secretary Brown's words, this calls for "calibrating U.S. retaliation to the provocation. . . . No potential enemy should labor under the illusion that he could expect to disable portions of our nuclear forces without in turn losing assets essential to his own military and political security, even if the exchange were to stop short of an all-out destruction of cities and industry."[22]

ALCMs contribute to assured destruction but the significance of the contribution is unpredictable. They do, of course, offer a large number of warheads that could strike industrial or population concentrations, but this will not represent an increase soon because they will for the most part replace other weapons (SRAMs and gravity bombs) now carried by bombers. By allowing B-52s to launch weapons from beyond Soviet air defenses, cruise missiles compensate for gradually deteriorating confidence in bomber penetrativity and a decline in the efficacy of the bomber leg of the triad of strategic forces.

As long as the submarine leg of the triad remains secure, however, the other legs are not critical to maintaining assured destruction capacity. And even if U.S. strategic forces were not modernized, bombers and ICBMs would retain a significant potential for simple retaliation against cities or factories.[23] Submarine-launched ballistic missiles (SLBMs) can cover a large number of "soft" targets, and as long as the Soviet Union does not deploy large numbers of antiballistic missiles, there is more assurance that SLBMs will penetrate than that air-breathing systems will. SLBMs should therefore logically have the primary responsibility for assured destruction. Over the long term cruise missiles offer a hedge against a Soviet technological breakthrough in antisubmarine warfare or an ABM breakout, but otherwise they are almost redundant for assured destruction purposes.

The most obvious contribution of cruise missiles to U.S. force posture is in essential equivalence. In the near future, at least, the USSR will not have cruise missiles comparable to American models. By virtue of their newness, the manner in which they diversify U.S. modes of delivery, and their potential capabilities for destroying hard targets (the limits of which are explored below), cruise missiles should help to offset the

22. *Department of Defense Annual Report, Fiscal Year 1981*, pp. 66–67.

23. The penetrativity of B-52s will probably decline only incrementally during most of the decade. Even if U.S. ICBMs are 95 percent vulnerable, 50 surviving Minuteman II and III missiles (with one and three warheads apiece, respectively) could deliver about 100 warheads (with a combined equivalent of roughly 100 megatons)—an impressive contribution to simple countervalue target coverage.

apparent gains in the strategic balance that its massive advantage in ICBM throw weight gives the Soviet Union. At a minimum, deploying ALCMs will moderate the perception of U.S. strategic inferiority in the mid-1980s by preserving the role of the aging bomber leg of the triad.[24] At a maximum, ALCMs will encourage the perception of a net increase in capability that will help plug the "window of vulnerability"[25] until more impressive counterforce capabilities (MX and D-5 ballistic missiles) become available. (Even if the ALCM is not a satisfactory quick fix for the window, it is the *only* fix aside from the Trident I C-4 SLBM currently programmed that will be available for the next few years.) Whether or not such perceptions are warranted, they should be intensified by the Soviets' vigorous complaints about the new threat posed by cruise missiles. And if cruise missiles prompt the USSR to invest more heavily in air defenses, they may reduce Soviet allocations to offensive systems, thus strengthening the U.S. position in the more visible elements of the balance.

The contribution of cruise missiles to a countervailing strategy is the most dubious. Their flexibility and accuracy are compatible with such a strategy in principle,[26] but not in practice unless (which is unlikely)

24. This was the principal consideration that led to earlier favorable assessment of ALCMs. See Alton H. Quanbeck and Archie L. Wood, with the assistance of Louisa Thoron, *Modernizing the Strategic Bomber Force: Why and How* (Brookings Institution, 1976). This study's conclusion that ALCMs could *substitute* for a new penetrating bomber (the B-1) by extending the efficacy of B-52s (or by allowing less capable carrier aircraft that could stand off beyond Soviet air defenses to perform similar missions) was controversial. See Francis P. Hoeber, *Slow to Take Offense: Bombers, Cruise Missiles, and Prudent Deterrence* (Georgetown University, Center for Strategic and International Studies, 1977); and Elmo R. Zumwalt, "An Assessment of the Bomber-Cruise Missile Controversy," *International Security*, vol. 2 (Summer 1977), pp. 47–58.

25. This catchphrase refers to the period, projected to last until more survivable U.S. ICBMs are deployed, in which land-based missiles are considered highly vulnerable to a Soviet first strike.

26. Accuracy permits the use of low-yield warheads, which reduce collateral damage and thus theoretically increase the capacity to wage nuclear war while holding significant portions of enemy population and economic assets hostage, thus deterring strikes against U.S. population centers. (The proximity of many military targets to urban centers, however, would make it extremely difficult to maintain these distinctions.) Flexibility is theoretically enhanced by the capacity to use some ALCMs for defense suppression and, with rotary rack launchers, to select specifically targeted missiles at will, rather than shooting them all at once or in a predetermined sequence. Moreover, storage of ALCMs mounted on pylons allows raising the alert level of bombers much more quickly than if they had to be loaded individually. *Hearings on Military Posture and H.R. 1872 [H.R. 4040]: Department of Defense Authorization for Appropriations for Fiscal Year 1980*, Hearings before the House Committee on Armed Services, 96 Cong. 1 sess. (GPO, 1979), pt. 3, bk. 1, pp. 646–49.

bombers can be withheld for a substantial time during war and the mission planning facilities for retargeting ALCMs remain intact. If plans called for using ALCMs very early in the nuclear phase of conflict, allowing the withholding of greater numbers of ballistic missiles, this could support a long-war strategy. But the demand for weapons capable of great accuracy and speed in response (survivable ICBMs would be the best such weapons) is likely to be highest in that initial phase.[27] Moreover, the counterforce capability supposedly offered by ALCMs is mitigated by their slow flight to target. The Soviets would theoretically have ample time to decide to discharge their reserve ICBMs before the ALCMs arrived. Thus the primary rationales for targeting cruise missiles against ICBMs would be to preclude the reloading of empty Soviet silos for a second salvo and, for purposes of deterrence, to deny the USSR the option of withholding substantial ensiloed reserves.

The principal question about the ALCM's capability is its potential to destroy hardened targets. For a long time cruise missiles were considered by both their strongest advocates and their most determined opponents to be excellent counterforce weapons. A decline in estimated ALCM accuracy has tempered these expectations. Calculations by Bruce Bennett and James Foster, which factor in numerous operational uncertainties, demonstrate that high-confidence effectiveness against very hard targets will require a warhead yield of 200 kilotons and missile CEP of 200 feet. Figuring an arrival probability of 80 percent, they conclude that an alert force of 1,350 ALCMs would be needed to damage 70 percent of Soviet silos (2,650 to damage 90 percent).[28] The potential

27. In 1977 one Navy analyst developed an argument for strategic employment of cruise missiles that presents them as superior to ICBMs with MIRVs for the requirements consistent with the strategy implied by public reports of Presidential Directive 59: "By selectively retaining spatially separated systems for reserve nuclear forces, the Soviets can widen the required footprint for the average MIRV, leading to a 'virtual attrition' of yield through increased fuel requirements or warhead wastage because of multiple expenditure at isolated targets. The attempt to fire at a nonprogrammed target complex further reduces flexibility for MIRV in that it may preclude optimum warhead-target assignment. The ability to dedicate a single warhead to a single target . . . represents flexibility for the cruise missile." But in a long war managing area queues of targets—that is, the gradual collection of targets located so that none of the warheads are wasted (which is more vital when retaliating forces have been degraded by preemption or drawn down by early operations)—is more complex. Either some of the MIRVs must be wasted in order to strike time-sensitive targets or firing must be postponed until the queue is filled. Edward J. Ohlert, "Strategic Deterrence and the Cruise Missile," *Naval War College Review*, vol. 30 (Winter 1978), pp. 27–28.

28. Bruce Bennett and James Foster, "Strategic Retaliation against the Soviet Homeland," in Betts, ed., *Cruise Missiles*, pp. 151–55, 158–63.

for effective target coverage by currently planned numbers of ALCMs appears even more limited in light of Presidential Directive 59, President Carter's 1980 guidance on nuclear strategy (which has reportedly been for the most part retained by the Reagan administration). The directive listed the types of targets that U.S. forces should be capable of attacking; the list included many hardened sites (such as command centers) besides silos—a total of about 2,700 hard targets.[29]

As long as U.S. war plans stipulate significant counterforce second strikes without launching the Minuteman force on warning, the ALCM's contribution will rest on one or more of the following: (1) a presumption of strategic warning, which would allow most of the bomber force to be on alert before Soviet attack (efficient queuing and flyaway of large numbers of aircraft would also be necessary); (2) some scheme for preferential targeting of ALCMs against the highest priority Soviet silos; and (3) reduction of the ALCM's CEP to improve the probability of destruction. The first assumption is risky, and the second places a high premium on correctly guessing Soviet operational priorities and estimating the risks to U.S. targets from different combinations of surviving Soviet ICBMs.[30]

Ensuring high ALCM accuracy would be the simplest and most direct improvement, though it would not offer a full "solution." Consideration might be given to fitting the missiles with the DSMAC system designed for conventionally armed variants. Reducing the CEP to a few meters could also allow lower-yield warheads, which could be tested within the threshold test ban limit of 150 kilotons and which would produce less

29. Ibid, pp. 146–47.

30. One rationale might be to concentrate on missiles apparently intended for military targets—SS-17s, -18s, and -19s with MIRVs. The disadvantage would be that most of these might already have been fired in the Soviet first strike and the ALCM force would have been used solely to prevent Soviet reloading. (One theoretical alternative would be to develop real-time surveillance and communication capabilities that could function under nuclear combat conditions and that, together with in-flight ALCM retargeting capability, could adjust cruise missile strikes to the pattern of Soviet first-strike launches of ICBMs. Such capabilities would not only challenge technical innovators, but necessitate high confidence in the smooth, integrated functioning of crews and equipment in unprecedented conditions of wartime stress and confusion.) Moreover, if SS-11s are replaced by larger missiles, within a few years *all* Soviet ICBMs may be "high priority." Ignoring this, one might search for a theoretical advantage: sparing the ICBMs that are part of the Soviet assured destruction reserve might make it less likely that the U.S. second strike would prompt the Russians to launch that reserve and destroy American cities. Such judgments are so sensitive to technical evolution—and so dependent on guesses about Soviet operational doctrine and reactions under attack and on refined game theory about coercion and escalation control—that they would be a perilous basis for planning. *Any* scheme for second-strike counterforce, however, runs many of these risks.

collateral damage. Doubts about the DSMAC's cost-effectiveness for strikes with conventional munitions would clearly not apply for strategic missions. The DSMAC, however, is sensitive to time of day and seasonal changes in the target area; thus plans for conventional missions rely on prestrike reconnaissance for digital updates. Though theoretically possible, this might not be practical for the thousands of strategic Soviet targets. Another hypothetical way to increase ALCM accuracy might be the use of global positioning system satellites. But even if practical (which it could not be before the end of this decade), this option would increase the ALCM's dependence on sophisticated support systems, an undesirable condition in a nuclear war. Finally, the CEP of the current models might be reduced by placing the missile's last TERCOM update map close to the silo, although this might increase its vulnerability to jamming. It is important to remember that, although the counterforce capability of the ALCM is most sensitive to accuracy, improved CEP alone would not compensate fully for other factors such as prelaunch survivability, reliability, and penetrativity.

It is apparent that no neat solution exists for increasing confidence in the cruise missile's ability to destroy hard military targets. Even so, despite Foster and Bennett's debunking of the high expectations of the ALCM's counterforce capability, they admit that it may offer better hard-target destruction potential than the other legs of the triad when survivability is considered. At this point the question of probable cruise missile attrition from air defenses becomes central.

Gordon MacDonald, Jack Ruina, and Mark Balaschak offer a confident view of the prospects for cruise missile penetrativity against large area targets. A devil's advocate with more pessimistic instincts might still find grounds for anxiety about Soviet air defenses. Official testimony leaves some doubt. According to former Under Secretary of Defense William J. Perry, the Soviet SA-10 could "have a formidable capability against the cruise missile," and by the late 1980s "it will be imperative for us to find the suitable countermeasures."[31] "Corridor cutting" by ballistic missiles to suppress defenses along cruise missile flight paths is possible, but this is a questionable comfort. In a second strike the United States might have few surviving ICBMs left, and SLBMs either would be husbanded as a reserve force or would have other targets to cover. And unless each ballistic reentry vehicle used against an air defense installation could clear the way for a significantly larger number of cruise

31. *Hearings on Military Posture and H.R. 1872 [H.R. 4040]*, pt. 3, bk. 1, p. 211.

missiles, corridor cutting would not be cost-effective: why use a warhead with 100 percent penetrativity merely to assist a cruise missile that is less certain to arrive? Moreover, if Soviet SA-10s are mobile, it might be impossible to plan reliable defense suppression. Nevertheless, vaunted Soviet progress in air defense would have to be far greater than the incremental evolutionary improvements of past decades to raise major doubts about the ALCM's efficacy against a target the size of a city. And if the stunning success of Israel's American-supplied aircraft in overcoming Syria's Soviet-supplied air defenses in Lebanon in June 1982 is any indicator of the efficacy of Soviet air defenses in general, there would seem to be less to worry about.

That the cruise missile would have greater difficulty in attacking hardened point targets protected by terminal defenses is widely recognized. Such defenses can be saturated, but only at the price of reducing the number of targets that can be covered, already a problem if ALCMs are allocated against ICBM silos. If dense defenses are deployed around the silos, the efficacy of ALCMs against those targets becomes quite dubious.

There remains the argument that U.S. cruise missiles will help restrain Soviet offensive capabilities by diverting additional resources into air defense investments. While the argument is valid to some extent, the assessment of the USSR's costs is overstated if "no allowance is made for the measures that are planned or underway in any case."[32] And if the extra investments changed the odds on penetration (as defenses around silos might, though increments to city defenses would not), the cost–benefit ratio might not favor the United States.

Despite all the limitations and uncertainties, it would still be wrong to conclude that cruise missiles have no utility for counterforce targeting. Just because they cannot satisfactorily cover all sites in a category does not mean they should not be used to cover some, especially if there are similar or greater limitations in the capabilities of alternate systems. Cruise missiles are only one element of U.S. strategic forces and need not accomplish all missions on their own. More complete appreciation of the potential contributions of cruise missiles depends on the overall U.S. strategy for deterrence and war.

32. Ron Huisken, *The Cruise Missile and Arms Control,* Canberra Papers on Strategy and Defence 20 (Canberra: Australian National University, Strategic and Defence Studies Centre, 1980), p. 13. Huisken estimates that if costs are spread over ten years "a dedicated Soviet defensive response . . . could absorb $6–7 billion a year."

Strategic Forces: Doctrine and Dilemmas

The credibility of deterrence depends in part on the credibility of war plans: what the adversary believes a victim could and would do with surviving forces if attacked. The drift of U.S. deterrence doctrine toward greater emphasis on counterforce options and targeting selectivity has been motivated not only by the changing strategic balance, but also by changing beliefs about Soviet conceptions of nuclear war. Given the durability of simple U.S. assured destruction capabilities over the coming decade, the magnitude of the ALCM's role depends largely on its value for striking military targets. To some extent the ALCM's efficacy against silos or other small-area targets depends on the scale of terminal defenses. If the Soviets perceive the ALCM as a worrisome threat, they may deploy more such defenses, reducing the attraction of using ALCMs against such targets. If the silos are only lightly defended against ALCM deployments, however, this suggests either that Soviet planners do not consider the silos especially vulnerable to cruise missiles or place little value on reloadimg them (in which case the deterrent logic of allocating ALCMs that way falls apart) or that they plan to launch on warning and rely on the reserve in surviving submarines.

The larger question is whether it is wise to threaten Soviet ICBMs even under the logic of the countervailing strategy. The rationale for doing so is rather convoluted.[33] Escalation control might argue against a threat large enough to provoke launch on warning. One argument is that

> even if Soviet missiles held in reserve as a form of blackmail against unattacked U.S cities were attacked by cruise missiles, the Soviet leadership might prefer to ride out the attack. This would leave a depleted but still existent blackmail capability of launching such missiles. A premature launch would remove the blackmail capability and result in a U.S. response against Soviet cities.[34]

33. Promoting a silo-destroying MX, General Alton Slay said of a Soviet decisionmaker faced with the missile: "If he attacks at this time he will say to himself the United States will still have the capability to attack his residual ICBM forces and countervalue targets. . . . He would then have to use his residual forces or lose them. That would escalate the conflict and invite further retaliation . . . in this case he would decide that an attack probably is not worth it. . . . Our strategic policy and programs ought to work on his mind and not on the silos. If we have to work on the silos, the policy has failed . . . [but] in order to work on his mind, we want to be able to work on the silos." Quoted in *Defense/Space Daily,* November 21, 1978, p. 96. As some proponents of assured destruction have argued for the "rationality of irrationality," this appears to argue for the "stability of instability."

34. Jeffrey T. Richelson, "Evaluating the Strategic Balance," *American Journal of Political Science,* vol. 24 (November 1980), p. 790, note.

This, however, suggests that if the Soviet "ride out" option is to be reasonable, the U.S. counterforce threat should be limited. In this case the modest ALCM counterforce capabilities in Bennett and Foster's assessment may be an advantage rather than an inadequacy.[35] In any case the uncertainties suggest that the public explanation of how ALCMs will be targeted should be left ambiguous.

One pair of arguments is that the ALCM is not useful for striking ICBMs because it is too slow, and that this slowness and the consequent lack of first-strike capability make the weapon stabilizing. Neither argument is fully persuasive. It is doubtful that Soviet leaders who have decided to gamble on a first strike would be less prepared to launch their reserve ICBMs upon short warning of incoming retaliatory ballistic missiles than upon longer warning of approaching cruise missiles. As for the presumption that cruise missiles are inherently second-strike weapons, Soviet leaders give no indication that they view cruise missiles as less dangerous than other, faster weapons; indeed, it is uncertain how far they accept the prevalent American conceptual distinction between first- and second-strike weapons.[36] A worst-case planner in Moscow

35. One possible contribution of cruise missiles to a deterrence-by-denial strategy could be rhetorical. Declaratory policy may be the only way to limit the perceived disadvantage in asymmetrical U.S. and Soviet capabilities (at least until the late 1980s) in the time-urgent counterforce capabilities unique to ICBMs. In theory, U.S. leaders could change course and emphasize the doubts about ICBM accuracy arising from the "bias" problem—the potential degradation of estimated accuracy by geophysical anomalies in the wartime flight paths of ICBMs, over which they have never been tested. They could also attribute optimistic CEPs to cruise missiles, which are not subject to the bias phenomenon (though such statements would have to ignore comparable uncertainties because TERCOM mapping was carried out on denied territory). Soviet leaders might then have less reason to feel confident that they had a net advantage in hard-target destruction capabilities. U.S. military leaders tend to dismiss the bias problem, but some scientists consider it significant. For an example of the former position, see the interview with Air Force Deputy Chief of Staff Kelly Burke in "Air Force's Chief Armorer Plans Sortie in Multi-Year Buying," *Defense Week,* February 2, 1981, pp. 4–7. For the latter view, see James Coates, "Some New Questions About the Accuracy of Missiles," *Chicago Tribune,* March 5, 1981; and Kosta Tsipis, "Precision and Accuracy," *Arms Control Today,* vol. 11 (May 1981), pp. 3–4. The administration's attitude about rectifying the strategic balance is generally more relaxed than that of President Reagan's principal preelection defense advisers, which suggests that a rhetorical ploy of this sort may not be politically infeasible

36. An official Soviet publication argues that a cruise missile cannot be detected when launched or in flight. "Consequently it is also a first-strike weapon like Pershing-2." Soviet Committee for European Security and Cooperation and Scientific Research Council on Peace and Disarmament, *The Threat to Europe* (Moscow: Progress Publishers, 1981), p. 24.

might postulate an inventive American attack plan using less effective but faster weapons (such as SLBMs) combined with European-based Pershing IIs (which are limited in number but accurate and even quicker) for "pin-down" strikes to disrupt the Soviet capability for response, and ICBMs and forward-based aircraft to destroy time-urgent targets, while the slower cruise missiles flew in to finish the job on remaining ICBM silos and command posts.

Bennett and Foster argue that the ALCM's contribution to evolving operational doctrine is not only modest but depends on the complementarity of other forces. The justification for a mixed force of cruise missiles and penetrating bombers extends beyond the need to stress Soviet air defenses, as long as the flexibility and endurance mandated by Presidential Directive 59 are criteria for the adequacy of strategic forces. The enduring reserve force would probably have to be constituted primarily of SLBMs. At least until a mobile ICBM is deployed, the flexibility is likely to have to be provided by the manned bombers. ALCMs would add to such flexibility indirectly, allowing bombers in the penetration phase to concentrate on in-flight damage assessment and restrike or (when and if reconstituted) to attack targets of opportunity or newly acquired targets.[37] Like much of the logic of Presidential Directive 59, this would require greatly improved survivability and efficiency of reconnaissance, command and communication channels, and transwar mission planning facilities.

ALCMs offer little for the element of the directive that has attracted most attention: the targeting of hardened Soviet strategic communications and leadership bunkers, which would logically be last (in time) on the list of military targets since striking them would be counterproductive before escalation control had failed and the war had become unlimited. The ALCM force could not easily be withheld beyond the first U.S. retaliatory counterstrike. The limitations of ALCMs under the criteria

37. Until the end of the 1970s U.S. strategic nuclear debate was fixated on the role of ICBMs with MIRVs. Some skeptics also interpreted Presidential Directive 59 as a cynical brief for the MX. However, the only novel operational emphasis implied in public reports of the directive—selective targeting executed over a long time span—are more conducive to single-warhead systems than to the ICBM with ten MIRVs. The advantage of a single-warhead cruise missile in avoiding warhead waste, though, is lost by its slowness, since target acquisition is most likely to seek movable assets in the later phase of war. These problems may be part of the reason behind official interest in developing a new, small, single-warhead ICBM. See Clarence A. Robinson, Jr., "U.S. Strategic Missile Options: U.S. Weighs Small ICBM Development," *Aviation Week and Space Technology*, May 4, 1981, pp. 49–52.

of countervailing strategy, however, are not exceptional when one considers the overall inadequacy of the strategic forces' support systems (especially command, control, and communications) and the lack of full commitment to the concept (much greater investment in civil defense would be consistent with plans for selective exchanges). As some observers have noted, the capabilities to implement the directive will remain an aspiration rather than a reality for quite a while. Thus cruise missiles are not an independent solution to doctrinal dilemmas, although they represent a useful partial improvement in force modernization, at least into the late 1980s.

Theater Nuclear Forces

According to current plans the United States will deploy nuclear-armed ground- and sea-launched cruise missiles. The GLCMs are to be placed in central Europe; sea-based TLAM-Ns can be used anywhere in the world. In both cases the political motives for deployment are clearer than the operational military rationales.

The principal impetus for the GLCM decision was the Soviet deployment of SS-20 missiles with MIRVs to replace older, less accurate, fixed-base, single-warhead SS-4s and -5s arrayed against Western Europe. Although GLCMs (in combination with Pershing II ballistic missiles) increase the quantity and quality of NATO's long-range theater nuclear forces, they are not a direct military counter to the SS-20s. Because the latter are mobile and have longer range (allowing them to be based beyond GLCM or Pershing striking distance), they cannot be reliably targeted by the new NATO systems.[38] While GLCMs cannot attack SS-20s, however, they dramatically increase U.S. capability to strike the western interior of the Soviet homeland without resort to central strategic forces.

The use of theater nuclear cruise missiles would probably be an intermediate step in escalation, between the use of battlefield tactical nuclear weapons and that of central strategic forces. Therefore, GLCM employment would probably be selective, to signal resolve and prompt Moscow to halt and negotiate but without inflicting enough damage to eliminate Soviet incentives for retaliatory restraint. Targets could include military installations far from population centers, tactical com-

38. See Roger H. Palin, "War in Europe: Nuclear and Conventional Options," in Betts, ed., *Cruise Missiles,* pp. 173–84.

mand and control centers for Soviet ground and air forces, supply depots, nuclear storage sites, railway yards, choke-points such as key bridges, railway yards, army and divisional headquarters, and rear assembly areas.[39]

At least two problems, however, are difficult to resolve. One is that, unlike ballistic missiles, which are invulnerable to current defenses and follow a trajectory rather than a zigzag flight path, cruise missiles would have to be launched in numbers sufficient to saturate defenses so as to assure target destruction, and those targets would not be immediately apparent to the Soviets even if the missiles' launch were detected. This would create a tension between military effectiveness and convincing Moscow that the strike was limited. A second problem is that the purpose of a selective strike with LRTNF is avoiding defeat without escalating to unlimited nuclear war. It is hard to see that this means anything but an attempt to restrict nuclear exchange to the theater. Yet Western European allies supported the decision to deploy GLCMs principally by the rationale that the deployment would reinforce the coupling of U.S. strategic forces to their defense. Such tension in the operational logic behind LRTNF modernization emphasizes the inextricability of military and political considerations in policy for theater cruise missiles.

SLCMs could perform much the same roles as GLCMs, and they have the advantage of global mobility. For theater nuclear missions they have some deficiencies. SLCMs could most easily be deployed, without inhibiting other missions of the delivery systems, on surface ships (although even on them, Tomahawks would compete for loading space with Harpoons, antisubmarine rockets, and antiair warfare missiles). Surface platforms, however, are vulnerable, and if they became nuclear delivery systems, they would also become higher-priority targets for the Soviet Union.[40] SLCMs on submarines are less vulnerable but pose other problems. Except for the ones to be mounted in new vertical launchers in ballast tanks, those deployed on existing attack submarines (SSNs) displace other weapon loadings (torpedoes). If only a few torpedo-tube-launched SLCMs are placed in each SSN, this penalty is not great. However, it does mean that only a modest number of SLCMs in this mode are available for theater strikes. In either case holding the

39. Ibid., p. 181.

40. There are doubts, though, that GLCMs will really be any less vulnerable. Even if dispersed upon warning, which is uncertain, the large and cumbersome convoys in which they travel would be hard to conceal over long time periods.

SSNs in firing position for land attack would probably require diverting them from their primary missions of escorting carrier task forces and hunting Soviet submarines. One naval officer sums up these doubts:

> Taking submarines, cruisers, and destroyers away from previously assigned tasks to move them to a launch area to await orders may be serious enough, but considering they will be in a more exposed position without task force defense in an area closer to the enemy's shore, the risks seem excessive for the few cruise missiles they may be able to deliver . . . [and] weapons load trade-offs may make these ships marginally ready for any mission but not adequately ready for their primary mission.[41]

The only clear way around these problems is to procure additional submarines and dedicate them to theater nuclear forces. This solution would be extremely expensive (a principal attraction of cruise missiles is their presumed low cost compared to that of other weapon systems) and unavailable in the near term (because of long lead times and trade-offs in shipbuilding); it also raises questions about manpower (the rigors of underwater patrol duty make it hard to field sufficient crews for existing submarines). Even if all these constraints could be handled, communications with submerged vessels are less flexible and reliable, which means that their timely response in selective strikes would be dubious.

If effectively deployed in some form, nuclear cruise missiles will enhance NATO capabilities directly by providing a larger number of warheads that can be delivered over great distances, in principle reducing the extent to which the use of theater weapons would destroy NATO's own territory in the process of striking Soviet forces. The indirect contributions could be equally significant. By allocating cruise missiles to fixed and heavily defended targets, many NATO dual-capable aircraft could be released for conventional missions in the critical first phase of battle. The relative worth of the military benefits offered by GLCMs or SLCMs, however, depends on the weight of their associated political and diplomatic costs.

Soviet Cruise Missiles—Threat or Relief?

Another question that bears on the security implications of U.S. investment in cruise missiles is the probable Soviet response. What would be the strategic consequences if the USSR decided to deploy

41. Johnson, "Tomahawk," p. 31.

ALCMs of its own? This issue has received scant attention in strategic discourse. Modern long-range missiles could make the intercontinental capability of the Backfire unambiguous. The Soviet Union has also reportedly been developing a new long-range bomber to replace its ancient Bears and Bisons.[42]

If Moscow's choice was between augmenting its forces with cruise missiles or not augmenting them at all, the United States obviously should view the prospect of Soviet cruise missiles with alarm. Conceivably this could happen if the Soviet decision was spurred by a desire to respond to the qualitative Western innovation rather than by cold strategic calculation of military requirements. In this case, if U.S. restraint could forestall a Soviet venture into a whole new dimension of weaponry, one might argue for more Western flexibility in the scale and timing of cruise missile deployments in the hope of reaching an arms control agreement that could satisfy U.S. and NATO requirements without driving the Russians into the cruise missile game. If one believes that Soviet decisions are determined primarily by careful strategic judgment, however, the desirability of cruise missiles will probably be measured against the apparent utility of alternate weapon choices and a decision will be subordinate to a more general determination of desired force levels. Soviet cruise missiles would then be less worrisome to the West.

Analysts who interpret U.S.-Soviet arms competition in terms of an "action-reaction" model might rate the chances of a Russian move toward cruise missiles as quite high. They would argue that a major self-defeating fault in some previous innovative U.S. programs (such as MIRV) was the failure to look ahead to the next stage, when Soviet imitation would neutralize the advantage or increase the threat. An action-reaction prediction, however, would not in itself be compelling. Consider the earlier difference in American and Soviet interest in old naval cruise missiles. The comparative lack of U.S. interest was due to the asymmetry of the two nations' naval missions and force structures. The reversal of interest when it came to modern long-range cruise missiles can also be understood in terms of larger trends and traditions in the organization of the countries' strategic forces. As the dominance of aircraft carriers discouraged early U.S. commitment to antiship cruise missiles, the dominance of the strategic rocket forces and ballistic

42. George C. Wilson, "Soviet Bomber Development Reported," *Washington Post,* June 27, 1979.

systems discouraged Soviet commitment to the air-breathing intercontinental delivery vehicles. The USSR did not emulate U.S. Air Force attachment to intercontinental bombers in its own strategic buildup.

Of course, the U.S. Navy now favors the Tomahawk antiship missile, and the Soviet Union might be driven toward ALCMs, not out of emulation but perhaps for the same reason as American strategists: the vulnerability of fixed ICBMs (although the USSR would be just as likely to respond, like the United States, by developing mobile ICBMs). There have, in fact, been some public indications of Soviet development of an ALCM.[43] Even inferior ALCMs on new Soviet bombers could be a credible threat against the continental United States, since U.S. air defenses are very weak.[44]

Does this mean that the United States should consider potential Soviet deployment of ALCMs a severe military danger? No. They would simply be a redundant threat in a first strike.[45] Indeed, the principal reason that Washington dismantled much of the U.S. air defense system is that the Russian ICBMs made the Soviet bomber threat only incidental. As long as the Soviet ICBM force remained substantial, ALCMs would not change that reasoning, unless it was assumed that increased air-breathing forces would offer options for a limited Soviet attack that would make investment in U.S. air defenses significantly more attractive. Washington should certainly prefer Soviet ALCMs to additional ballistic missiles. And if cruise missiles only buttress Soviet second-strike capabilities, so much the better for the stability of mutual deterrence. That, after all, is the popular justification for why the USSR should not see American slow-flying ALCMs as a threat to stability. It is true that Soviet cruise missile deployments might be less susceptible to U.S. intelligence monitoring. But the idea that "the threat we know" is preferable to one

43. When asked during Senate testimony whether the USSR had an ALCM in research and development, Chairman of the Joint Chiefs of Staff David Jones replied, "We have some very limited evidence that I would not like to discuss in open session." *Military Implications of the Treaty on the Limitation of Strategic Offensive Arms and Protocol Thereto (SALT II Treaty),* Hearings before the Senate Armed Services Committee, 96 Cong. 1 sess. (GPO, 1979), pt. 1, p. 294.

44. Huisken, *The Cruise Missile and Arms Control,* p. 51.

45. Some have suggested that blind spots in U.S. radar coverage could allow Soviet bombers to sneak through in a surprise strike. The risks to the Russians in such a plan, if it were not simply complementary to an ICBM attack, would be extreme. If it was complementary, the incremental threat to U.S. targets would be small. One scenario sometimes cited to justify concern is that the bombers could sneak through before ICBMs were launched, striking command and communications targets and bomber bases and preempting U.S. capacity to launch forces on warning.

we do not know is questionable if the technical capabilities of the former (modern large ballistic missiles) are more dangerous.

Some observers worry about the budgetary pressure that might result if Soviet cruise missiles prompted stronger interest in continental air defense. This might be significant politically if larger Soviet air-breathing forces highlighted the disparity in air defense, but it would not change the military reasons for Western limits on such defenses as long as vulnerability to Soviet ballistic missiles is so great. The cost-benefit attractiveness of switching U.S. resources back to air defense would vary directly with the extent to which Soviet ALCM deployments *replaced* more worrisome ICBMs. Fear of the USSR's investment in cruise missiles would rest on the assumption that its alternatives were either adding ALCMs and increasing the threat or keeping the threat at the present level. It is more likely that the choice would be between adding cruise missiles or adding ICBMs; the latter would be worse for the United States. Reports that Moscow has considered developing more modern SLCMs to deploy off U.S. coasts in response to GLCMs in Western Europe are alarming,[46] since SLCMs might be able to sneak in under radar. But warning time of a Soviet SLBM attack on coastal targets (including Washington, D.C.) is already ten minutes or less—hardly enough to make a great difference. Although worrying, it is not clear that SLCMs would be more worrying than the deployment of more SLBMs, which might carry MIRVs.

Similarly, for theater nuclear deployments Soviet GLCMs would be less threatening to NATO than a comparable number of additional SS-20s. (By the same token, it is unclear what the military attractiveness of cruise missiles over SS-20s would be to Soviet planners. Unlike NATO, Moscow does not need the increased range of cruise missiles to broaden its coverage of European targets.) When the nuclear alternatives are considered—as either replacements or additions to force levels—it is reasonable to argue that in military terms the West should encourage a Soviet move from ballistic to cruise missiles. This felicitous conclusion, however, may not carry over to the realm of conventionally armed cruise missiles. If Moscow matches Western investments in such systems (already of tenuous potential cost-effectiveness), the improvements such systems promise for NATO's position in the conventional military balance may well be neutralized. More than the technically different

46. Ned Temko, "Moscow Developing New Cruise Missiles," *Christian Science Monitor*, April 13, 1982.

force elements in the strategic nuclear area, conventional cruise missiles offer as much for initiative and counterforce war-fighting as they do for retaliation and deterrence. If the Russians use conventional cruise missiles to threaten airfields in Western Europe and to interdict reinforcement choke-points more effectively, the United States will be just as hampered in prosecuting a war as the Soviet Union will. As one anonymous defense expert said, "If it's cheap for us, it's cheap for them."[47]

Conventional Strategy

Cruise missiles were originally developed to augment nuclear forces, but in recent years enthusiasm has grown for their potential use with conventional munitions. No clear demonstration has yet been made, however, of the cost-effectiveness of conventionally armed cruise missiles for any but a limited number of missions. While single-shot systems costing in the neighborhood of $1 million apiece are clearly worth expending to deliver highly destructive nuclear ordnance, the number of very valuable targets that could be destroyed by a conventional warhead of 1,000 pounds or less *and* that could not be effectively attacked by a more expensive but reusable manned aircraft is uncertain.

NATO Defense

The utility of conventionally armed cruise missiles varies principally according to two criteria: the desirable depth of interdiction strikes, and the priority given targets that are well defended and hence forbidding to manned aircraft. The first criterion places a progressively higher premium on the technical refinement of the cruise missile's aerodynamic and munition capabilities beyond those of the current generation, since the range-payload trade-off becomes more exacting as the depth of operation increases. The more difficult or expensive such refinements are, the narrower the range of advisable applications for conventional cruise missiles. Assumptions are often too blithely made about how appropriate the target arrays deep in Eastern Europe would be for conventional strikes. Some would be area targets, and some point

47. Quoted in David Fouquet, "Cruise Missile: Launcher for Next Arms-Control Talks?" *Christian Science Monitor*, December 17, 1979.

targets—especially bridges—are too hard for their destruction to be assured with less than perfect accuracy. Unless the missile actually hits a bridge pier, abutment, or span, the bridge will probably not go down. If raising the probability of destruction requires using a significant number of cruise missiles against the target, sending manned aircraft with precision-guided munitions to strike it becomes more attractive. Range constraints on manned aircraft would make cruise missiles preferable against such targets at long distances, but maximal missile range exacts a cost in the size of the explosive payload, which in turn further reduces the probability of destruction.

In central strategic force planning, improvement of Soviet air defenses would cast doubt on the comparative value of ALCMs and ballistic missiles. The same trade-off applies with even more force to tactical Soviet air defenses. Conventional cruise missiles become more attractive, however, if the alternative is reliance on manned aircraft rather than on rocket-powered tactical missiles. Less attrition would be likely if the aircraft were not required to attack heavily defended points, and they would also be freer to concentrate on targets for which the unique flexibility of a manned system is most useful: mobile forces or targets of opportunity. For missions such as striking airfields (by dispensing submunitions along a horizontal attack profile), conventionally armed ballistic missiles may not be competitive with cruise missiles or fighter-bombers in range-payload combinations and effectiveness per unit.

Cruise missiles are not immune to attrition en route, and the density of Soviet tactical air defenses presents a much greater barrier than simple extrapolation from strategic defenses in the Soviet homeland would suggest. If Soviet theater air defenses improve, there will be a stronger incentive to incorporate more agility and "intelligence"—for example, electronic countermeasures—into the missile. Thus the pressure to make conventional cruise missiles more sophisticated—and therefore less certainly a cost-effective investment—comes from all directions.

Simplified cost comparisons demonstrate how sensitive the value of first-generation conventional cruise missiles is to assumptions both about the number required to destroy a target and about manned aircraft attrition rates.[48] If estimates of aircraft losses are highly optimistic, airfield attack with very few cruise missiles would have to be feasible for missiles to be sensible for that purpose. If one considers the high Israeli loss rates during part of the October 1973 war to be grounds for

48. See John C. Baker, "Program Costs and Comparisons," in Betts, ed., *Cruise Missiles*, pp. 115–23.

expecting attrition of 7 percent or more, on the other hand, cruise missiles are certainly competitive with the option of purchasing additional aircraft—especially when related costs of pilot training and the inestimable value of pilots' lives are factored into the equation.

Cruise missile investments appear sensible at the theoretical attrition "break-even" point (where tactical aircraft costs and cruise missile costs for performing the same mission are equal), no matter what it is, because a mixed force provides more adaptability when neither element has a decisive advantage over the other. But if budget limitations required that such investments be balanced by significant *reductions* in manned aircraft (rather than simply adding different capabilities to force levels in the current structure), estimated aircraft losses would have to be discernibly higher than the break-even point to justify shifting reliance to cruise missiles. The value of manned systems' flexibility—to roam over or behind the battle area, strike mobile targets of opportunity, and cope with mobile defenses—is substantial. Cost-effectiveness methodology breaks down as a reliable basis for procurement decisions at this level, since the effect of intuition varies directly with the emphasis on flexibility, which by definition involves a large number of variables and subjective judgments not amenable to clear quantification. Systems analysis calculations offer sufficient grounds for reducing the number of aircraft in this case only to the extent that doctrine shifts toward more reliance on preplanned strikes and greater division of labor within the force structure and less emphasis on wartime adaptability in executing missions. Finally, aircraft are not the only competitors to cruise missiles. Other types of missiles, with rocket power, could be useful.

The balance of military considerations (this excludes arms control complications) seems to favor the deployment of conventionally armed cruise missiles.[49] But what should be their principal launch mode? The advantages offered by the MRASM are (1) higher explosive payload allowed by lower fuel requirements based on the carrier aircraft's ability to launch the missile closer to the target; (2) the avoidance of GLCM support costs, which would be greater than the incremental costs attributable to the aircraft from the missile; and (3) possibly greater compatibility with arms control verification schemes (aircraft may seem more readily countable than mobile or warehoused TELs). The principal advantage of a hypothetical conventional GLCM is the increased availability of aircraft (which otherwise would use flight time carrying the MRASM and would carry less ordnance of other types that could be

49. See ibid., and Palin, "War in Europe," pp. 184–96.

used during the balance of the flight) for missions that make the best use of their unique flexibility. SLCMs could be a possible complement, though perhaps only from surface platforms if previously mentioned problems of mission overload for attack submarines are a real consideration.

Flexibility is generally desirable because of the unpredictability of operations with a modern force structure untested in such a war as would occur between NATO and the Warsaw Pact countries. This argues for a mixed MRASM-GLCM-SLCM force. If a choice must be made or the relative emphasis between the variants be determined, decisions should hinge on two variables on which there is as yet insufficient evidence. The principal one is whether East-West arms control negotiations have any future and, if so, whether conventional GLCMs or SLCMs would be compatible with an agreement. The second is whether simple mechanisms for multiple reload of GLCMs, which would allow reliance on a small number of launchers and thus limit infrastructure costs, are feasible.

Roger Palin favors the GLCM,[50] but its potential for frustrating arms control is tremendous. The logical solution is to configure the missile with functionally related observable differences (FRODs) to differentiate it from the nuclear variant. FRODs, however, have been suggested in principle but not well defined in practice, and it is difficult to envision what they would be. (If future conventional cruise missiles are redesigned and tailored specifically for certain tasks, the differences may be persuasively "functional.") A more feasible alternative would be simple external observable differences (EODs)—any appurtenances not visible on the nuclear GLCM, even if the difference did not physically preclude a nuclear mission. While more feasible than FRODs, EODs would be less credible for verification but offer a better basis for bargaining than no observable differences at all. One form of EOD not integral to the missile but perhaps reasonably convincing would be the lesser security measures associated with GLCM convoys containing no nuclear material. Or, similar to a suggestion made by Palin, conventional GLCMs might be deployed in a fixed or limited mobile mode at air bases—no more vulnerable to conventional preemption than aircraft—while fully mobile nuclear GLCMs remained theoretically more survivable and thus more of a deterrent against Soviet nuclear preemption.

50. See ibid., pp. 199–205.

Outside the NATO Central Front

Ground-launched cruise missiles could be deployed in other regions, such as Korea, although there are no current plans to do this. In general, land-attack SLCMs or MRASMs should be preferable because they are more flexible—they could be moved quickly from one region and brought to bear in another as circumstances dictated. Although cruise missiles could perform missions around the globe, there are probably not many cases in which using them would be preferable to using other systems, such as manned aircraft.

Early in the Reagan administration plans were announced to recommission several *Iowa*-class battleships, and this decision was explicitly justified by the availability of cruise missiles to give the ships long-range firepower.[51] Even if such platforms for SLCMs are unusually survivable in conventional engagements (the battleships have extremely heavy armor), planning to use them as substitute aircraft carriers for penetrating Soviet home waters (for example, to strike the Kola Peninsula or Petropavlovsk) may be foolhardy. To get close enough to fire missiles inland, the ships would put themselves within the range of Backfire bombers, and this argues for protection by carriers, thus bringing the rationale for deploying the battleships full circle. If the crunch really came, even peacetime deterrent rhetoric about using carriers in these areas would probably evaporate; per unit they constitute too large a proportion of the service's striking power to risk in missions that could turn into a naval Charge of the Light Brigade. Planning to use the platforms to attack coastal rather than inland targets would avert interservice conflict over role assignments (the Air Force cedes only naval infrastructure targets on land) and would allow the ships to remain further back, though not enough to measurably decrease their vulnerability. Moreover, conventionally armed cruise missiles would be wasted on large target areas such as shipyards or hardened ones such as submarine pens.

Large surface platforms like battleships make more sense for contingencies in remote areas where aircraft carriers are unavailable, but the

51. According to Undersecretary of the Navy Robert J. Murray, "The battleship is returning because the new technology of cruise missiles . . . opens up a new role for the battleship." "Bring Back the Battleships," *Washington Post,* April 12, 1981. See also R. D. M. Furlong, "US Navy Bringing Back the Battleships?" *International Defense Review,* vol. 14, no. 6 (1981), pp. 741–45.

value of this option depends on the priority placed on intervention operations in third world regions like the Indian Ocean or Southeast Asia.[52] The narrow confines of the Persian Gulf make it a dubious candidate. If any ship with cruise missiles is to be put at risk there, it would probably be more sensible to use one or more of lower unit value, such as a destroyer. Launched from outside the Gulf, SLCMs could cover few of the targets relevant to most Middle Eastern or Iranian contingencies. MRASMs might reduce the limitations but would require a carrier rather than a battleship. Another alternative that has been suggested is to use land-based B-52Ds (scheduled to be retired) to cover some sea lanes. The Navy opposes this idea, perhaps because it encroaches on the rationale for expanding the fleet to 600 ships.

With the exception that constructing new platforms dedicated to launching cruise missiles would be more costly, it is unclear that the battleship option is justified. (And administration cost figures have not emphasized the daunting manpower requirements—over 1,500 for each battleship—that will be difficult to meet without conscription.) In low-threat environments weaker task forces might suffice, yet it is doubtful that the stronger battleship could make use of its unique mix of capabilities (such as the sixteen-inch guns that require close approach to the shore) in any other than low-threat environments. And in such "easy" areas the cruise missile's penetrativity against air defenses may also be superfluous and thus less cost-effective than alternatives. Even if aircraft carriers are unavailable, quick-reaction B-52s with old-fashioned iron bombs might do the job just as well. Indeed, the Strategic Air Command has been planning this sort of option.[53]

52. See Richard Burt, "Local Conflicts in the Third World," in Betts, ed., *Cruise Missiles,* pp. 213–29.

53. Plans reportedly call for using the B-52H because of that model's 8,000-mile range. From bases in Spain or, better, Egypt, they could fly missions to the Persian Gulf without refueling. The planes now assigned this mission have been christened the "Strategic Protection Force." Howard Silber, "B-52s Able to Speed to Mideast Hot Spots," *Omaha World Herald,* January 25, 1981; and George C. Wilson, " 'Anytime Anywhere': A New Conventional Role for B52 Bombers," *Washington Post,* March 1, 1981. The strategic mission makes the H-model a dual-capable aircraft. If the former report is accurate, however, SAC planners must consider the political fact that Spain (with close ties to Arab states) or Egypt could easily forbid the use of the bases. Moreover, the use of the B-52H for conventional strikes would probably have to be limited to contingencies that did not involve probable combat with Soviet forces; otherwise those aircraft might be withheld for SIOP (single integrated operations plan) missions. If not, and if SAC can afford to take a number of the best B-52s out of the SIOP, the need for a new bomber seems harder to justify.

The role of cruise missiles in war at sea raises a number of different issues.[54] The Tomahawk antiship missile's principal significance is that it offers a new means of attacking ships at long range. One promising option not yet widely discussed is the deployment of *shore-based* TASMs astride crucial coastlines such as the Norwegian Sea, the Mediterranean, and the Sea of Japan to capitalize on Western geographic advantages. Reaching out 250 nautical miles, these TASMs could cover crucial transit points or operating areas of Soviet naval forces.

Ship-based TASMs could be used against Soviet surface fleets, but their utility in this role has some drawbacks. First, TASMs cannot assuredly preempt the Soviet use of naval cruise missiles. TASMs might sink opposing vessels but—because of their flight time—not before those vessels launched their own missiles. Second, large numbers of TASMs would have to be used against major surface formations since large ships would probably not be disabled by a few hits, and more missiles than would actually strike home would have to be launched to saturate air defenses.[55] (The sinking of H.M.S. *Sheffield* by an Argentinean Exocet missile during the 1982 war in the South Atlantic raises some questions about this conclusion, but the *Sheffield* was a small ship in an isolated position without appropriate antimissile defenses such as the Sea Wolf system.) Over-the-horizon target location is another problem, especially if many land- and space-based surveillance systems were destroyed early in the war. Overall, however, by releasing carrier aircraft for other missions (such as fleet air defense), TASMs and land-attack SLCMs may offer a significant net increase in naval capabilities against the Soviet Union.

Politics and Diplomacy

Procurement and changes in force structure should logically be consistent with military operational doctrine and the broad goals of strategy. These are necessary though not sufficient conditions for judging how and to what extent cruise missiles should be integrated into the U.S. defense posture. The cruise missile's multipurpose nature makes the third group of variables—political factors—dominant. The analyst's

54. See Michael MccGwire, "The Tomahawk and General Purpose Naval Forces," in Betts, ed., *Cruise Missiles*, pp. 231–73.

55. For example, MccGwire estimates that about eighteen TASMs might have to be launched to disable the *Kirov*. Ibid., pp. 238–39.

temptation to address problems from the perspective of a "Strategic Man" must be subordinated to the realities of political disagreement and competition where budgets are finite and negotiations or adversary and allied reactions affect the balance of power as much as unilateral initiatives do.

The political aspects most easily controlled by American leaders are internal. In contrast to weapon programs such as the MX, there is negligible controversy about cruise missiles—and thus no appreciable basis for constraint—in the body politic. Only the intragovernmental politics of legislative-executive relations, organizational interests, and bureaucratic bargaining apply. The origin and evolution of cruise missile programs were sensitive to such factors. Much of the gestation of the cruise missile, however, is past. Henceforth, bureaucratic politics are likely to be a less significant determinant of policy for cruise missiles (compared with external factors) than for other weapons that are not so formidably enmeshed in alliance politics and U.S.-Soviet relations.

These last two factors could compromise the logic of strategic calculation. If serious nuclear arms control that could limit Soviet forces remains possible, some conventional applications of cruise missiles may have to be sacrificed, their relative contribution to theater nuclear forces may be scaled down, and potential interest in proliferating ALCMs beyond levels planned under SALT II could be squelched. Yet if arms control dies the United States will not be free to capitalize fully on cruise missile deployments. Although allies pushed commitment toward these weapon systems in the late 1970s, they could derail the programs in the early 1980s. (The 1979 NATO decision on LRTNF was premised on ratification of SALT II and conduct of theater nuclear arms control talks.) Any interment of SALT or theater nuclear forces negotiations for which Washington appears to bear more responsibility than Moscow could demolish the fragile 1979 plan for GLCM deployment and might fracture the Atlantic Alliance. Policy cannot be made with the coherence or decisiveness of unitary Strategic Man in an alliance of sovereign states with varying government views, priorities, propensities to take risks, and domestic constraints.

American Defense Planning

Professional military demands for cruise missiles or resistance to them is determined in part by objective assessments of their utility, in part by the influence of doctrinal predispositions under conditions of uncertainty in those assessments, and in part by anticipatory suspicions about the effect of innovative programs on other valued elements of force structure under conditions of budgetary stringency. As long as a new weapon appears to be "free" (unthreatening to other programs and traditional attachments to doctrine), it will be welcomed enthusiastically.

In the early 1970s defense budgets were declining and cruise missiles seemed a cheap alternative for executing some missions. Service support for the weapon was limited, conditional, or nonexistent. Attack submariners liked it in principle but opposed it because of the competition for dollars and the pressure to trade off such new weapon loadings against others. Chief of Naval Operations Elmo R. Zumwalt, Jr. (a surface sailor), liked it in principle but opposed it as a justification for new submarines. Similarly, through the late 1970s the ALCM was promoted as a substitute for penetrating bombers, so Air Force officers opposed it. Throughout the decade the Pentagon's push for cruise missiles came from the Office of the Secretary of Defense rather than the military.

By the beginning of the Reagan administration the defense budget was growing, the issue of substantial commitment to cruise missiles was closer to being settled, and eventual deployment of some new manned bomber seemed assured. Although it is now evident that the defense budget increases originally envisioned by Reagan will be scaled down, substantial growth is still a certainty. While cruise missiles may be bought at the price of *additions* to other elements of force structure, significant offsetting reductions are unlikely. Service resistance, therefore, probably figured more in the genesis of cruise missile programs than it will in their fruition. Helped along by "commercial push" (industrious salesmanship by cruise missile manufacturers who have touted the system as a panacea, wonderful for attacking everything from tanks to submarines) and by "conventional push" (multiplying plans for tactical applications reinforce the general level of interest in the basic system), momentum now characterizes the weapon's progress more than sluggishness. However, economic pressure will make trade-offs between programs more exacting. The momentum of cruise missile programs also raises the possibility of unforeseen diplomatic ramifications and effects on Soviet defense policy reactions.

In this context the Office of the Secretary of Defense may act more as a restraint (or arbiter) than as the prod it was in the past decade. Otherwise, if the administration relies heavily on recommendations from military staffs instead of judgments by units such as the Office of Program Analysis and Evaluation, investment in different cruise missile variants may follow a general pattern of aggregation—procuring every variety favored by the three services, with numbers arbitrarily limited according to the incremental funds available—rather than discrimination according to cost-effectiveness analyses that cut across service lines. Historically, weapon innovations were often incorporated in force structures without appropriate doctrinal adjustments because of the inadequacy of organizational arrangements for evaluation.[56] The American military does not have a genuine General Staff, of the sort for which the German tradition is renowned, to deal trenchantly with mission and force structure trade-offs across service lines. Without fully integrated programming decisions, ambiguities in the assessment of cruise missile utility will compete with uncertainties in the achievable levels of procurement for other systems.

This may not matter if political complications or strategic utility are primarily sensitive to the overall level of conventional cruise missile deployment rather than to the specific variations within that level. And in any case the significance for future policy outcomes of the bureaucratic decisionmaking process is likely to be marginal. The important political constraints are not intradepartmental staffing procedures, but the external pressures that could affect interagency discussion and presidential decisions.

56. I. B. Holley, Jr., *Ideas and Weapons: Exploitation of the Aerial Weapon by the United States during World War I* (Yale University Press, 1953), pp. 16–17, 19–22, 176. Development of early cruise missiles in the 1950s was hampered by a management system that dealt with them as a simple extension of aircraft research and development; this did not optimize the use of subsystem technologies. In contrast, ballistic missiles, which were seen as "different," were handled in a more flexible, project-centered organization, which allowed "high-level intervention to lift their development out of standard organization." Robert F. Coulam, *Illusions of Choice: The F-111 and the Problem of Weapons Acquisition Reform* (Princeton University Press, 1977), pp. 354–59. In the early period, "development of the cruise missile as a weapons system did not keep pace with the evolution of its component parts." The present ALCM and SLCM programs, however, "began in 'crisis' periods during which normal weapon development routines were disrupted. This permitted extraservice organizations with interests in promoting cruise missiles and their associated technology to exercise important influence. A new 'action-channel' was momentarily created and exploited to achieve a reorientation of ongoing service-sponsored programs." Henry D. Levine, "All Things to All Men: The Politics of Cruise Missile Development," *Public Policy*, vol. 25 (Winter 1977), pp. 125, 122.

Crosscurrents in Arms Control

Even before the 1980 election the future of SALT was in doubt. The rhetoric of the Reagan campaign provided still stronger grounds for doubting that the negotiations themselves, let alone the prospects for a viable treaty, would survive. Further deterioration of U.S.-Soviet relations since the Polish crisis leaves little hope that the situation will change. But now that the president's task has shifted from criticizing past arms control agreements to finding ways to cope with the prospective as well as the present Soviet nuclear threat, START (strategic arms reduction talks, the new name for SALT) should remain an attractive option in principle. Pressure from U.S. allies and the desire to prevent LRTNF modernization plans from falling apart also encourage serious pursuit of arms control. The formidable challenge for future negotiation is to devise formulas that make more ambitious criteria for the adequacy of American nuclear forces compatible with the Soviet Union's minimum criteria for an acceptable balance. Otherwise, the Soviet leaders' greater domestic political freedom of maneuver may compensate for their inferior productive potential and allow them to make additional military investments that will neutralize American improvements.

Cruise missiles are a secondary element in this problem where intercontinental forces are concerned (for theater nuclear forces, missiles are more important). Soviet ICBM development is the principal concern of Washington, and the future MX is therefore the responsive system on which most attention is focused; since the availability of carriers for ALCMs is limited, restrictions akin to those in the SALT II treaty—which counted ALCM carriers under the permitted number of so-called MIRVed launchers—are unlikely to be especially problematic until much later in this decade. The GLCM is another matter, since it brings strong Soviet and Western European sensitivities into the picture, but this involves the linkage between START and theater nuclear arms control.

It is ironic that until the last few years the arms control advocates most firmly committed to institutionalizing a situation of mutual assured destruction saw ALCMs as a major threat to stability. The ALCM appeared to derail the Vladivostok agreement, augur a proliferation of U.S. weapons that would drive up force levels on both sides, destabilize the balance by dramatically increasing U.S. counterforce capability, and frustrate verification. But SALT II managed to produce a formula for verification, and degradation of early popular estimates of the ALCM's hard-target destruction potential combined with the missile's association (in the Carter administration) with the cancellation of the B-1 bomber to make it appear more compatible with mutual assured destruction theory's norms of arms restraint, while the evolving doctrinal trends highlighted by Presidential Directive 59 made the MX a much more salient issue.

Cruise missiles became less problematic in the restricted frame of reference of SALT I and II negotiations (intercontinental bombers and missiles and SLBMs), but the frame of reference itself has become more problematic, in part because of the cruise missile's prospective role in theater nuclear forces. This highlights the question of whether the difficulties for negotiated limitations posed by cruise missiles would be outweighed by the contribution to deterrence the weapons may offer through unilateral adjustments in force posture.[57] Is formally regulated arms race stability necessarily preferable to a sort of unregulated "free market" stability, which might require giving up the goal of leveling off arms competition and reducing innovation and procurement on both sides? Negotiated arms control does not ipso facto create greater deterrence stability than competitive innovation and deployment do.

Even if the negotiating problem could be kept within the old SALT frame of reference, the cruise missile presents challenges for the future: the potential for proliferating ALCMs is growing, and the system will be linked with the ICBM problem (implicit in the sublimit counting rules for MIRVs of the SALT II treaty) even if Washington manages to solve the MX basing problem and begins to deploy the missile, and even more so if it does not. For the long term, both the significance of the ALCM for maintenance of the American deterrent and its relative importance in arms control negotiations will vary with the role of the MX. If the MX program proceeds, the United States will have less incentive to push larger numbers of ALCMs into the strategic force, and they will therefore be a less significant part of the Soviet concern about U.S. capabilities. If U.S. ICBM modernization remains frustrated, on the other hand, there will be temptations to temporize by augmenting the other legs of the triad. Since the possibility for extra submarine construction is less

57. See George H. Quester, "Arms Control: Toward Informal Solutions," in Betts, ed., *Cruise Missiles*, pp. 275–307.

elastic than that for cruise missile carrier aircraft, boosting the number of ALCMs would be a more likely option.

If there is a future in START, innovative formulas may be necessary. One possibility, for example, might be to tie the average number of ALCMs on U.S. aircraft to the number of MIRVs on Soviet ICBMs,[58] especially if the ICBM element in the U.S. force is degraded by the inability to solve problems with unilateral force improvements. In that event, though, it would be all the more vital to hold out for a differential in the comparison—allowing a higher average of ALCMs per delivery vehicle than Soviet MIRVs—because air-breathing missiles are less penetrative than ballistic reentry vehicles. As long as the ABM treaty remains in force, Moscow can hardly deny the logic of a differential without denying the utility of the Soviet air defense system (and denial would give Washington grounds for a demand to limit such air defense deployments).

This rationale also helps meet U.S. concern about the greater threat posed by "time-urgent" counterforce capabilities of Soviet ICBMs without resting on the distinction—which Soviet spokesmen do not accept—that slow bombers and cruise missiles are only retaliatory weapons without first-strike utility. Moreover, if a survivable MX deployment does go forward and Soviet negotiators are more worried (as they logically should be) about that system than about ALCMs, a differential average for trade-off between ALCM carriers and ICBMs with MIRVs should be a reasonable compromise, mitigating Soviet anxiety about their vulnerability (given the alternative U.S. choice of emphasizing MX in the force mix) and U.S. qualms about bomber survivability and ALCM penetrativity. Such a formula, of course, does not solve many potential negotiating problems, and neither side is likely to welcome it, but both could see it as a second-best solution—and more feasible than alternatives.

Theater Arms Control

Outside the realm of central strategic forces as traditionally defined by Americans, the interlocking military, political, and diplomatic challenges posed by cruise missiles appear less tractable. Many U.S. arms

58. See William H. Kincade, "Arms Control: Negotiated Solutions," in ibid., pp. 325–26.

controllers prefer to keep theater cruise missiles nuclear, because significant deployments of conventional models would threaten negotiated agreement by muddying the verification problem. It is the nuclear cruise missiles, however, that provoke the most opposition on the Western European Left.[59] The whole problem of grappling with long-range theater weapons in arms control is relatively new compared to the more than ten-year history of SALT, and it raises the ambiguities and dilemmas in alliance relations more insistently than the earlier negotiations. In the 1950s Soviet deployments of medium- and intermediate-range missiles

> were more than offset by a substantial U.S. intercontinental superiority. So much so, that they were largely forgotten. After the beginning of MBFR [mutual and balanced force reductions], there even emerged, for a period, an image of massive Western superiority in theater nuclear forces stimulated by the artificial delimitation of the area to which MBFR is to apply, the so-called NATO Guidelines Area . . . [which] excludes the Soviet Union but includes the Federal Republic of Germany, and since the Soviets have been hesitant to deploy nuclear weapons beyond their borders.[60]

Indeed, the recent push to engage in theater nuclear arms control reverses the earlier alliance position, when Europeans preferred to keep U.S. theater nuclear forces out of SALT to avert a decoupling of U.S. strategic forces.[61] The reversal was not accomplished by resolution of the old concern, so European ambivalence toward theater nuclear forces has been heightened. Many European leaders want nuclear GLCMs to be strategically coupling, but they do not want them to unnerve the Soviet Union by being strategically threatening. In a sense, therefore, they want GLCMs to be strategic but not too strategic. This contradiction does not augur well for the possibility of comprehensive arms control that is already slim because of the fate of the SALT II treaty, Poland, and the vagaries of American domestic politics.

While Carter's opponents damned the SALT II treaty as a codification of U.S. weakness, many European leaders tended to see the potential rejection of the treaty as evidence of weakness in America's capacity for leadership.[62] The divergence in European and American views is

59. See Catherine McArdle Kelleher, "West Germany and the Nonnuclear NATO Countries," in ibid., pp. 481–508.

60. Stanley Sienkiewicz, "Foreign Policy and Theater Nuclear Force Planning," *Journal of Strategic Studies,* vol. 2 (May 1979), p. 28.

61. Lawrence Freedman, "The Dilemma of Theatre Nuclear Arms Control," *Survival,* vol. 23 (January–February 1981), p. 3.

62. John B. Ritch III and Alfred Friendly, Jr., *SALT and the NATO Allies,* Committee

often explained by Europeans' greater concern with the political and diplomatic elements in the problem of nuclear policy and Americans' greater concern with the technical and military elements. This simple distinction masks a deeper difference in dominant views of deterrence. It implies to some Americans that the problem is basically one of weak will, that European governments are more inclined to appease their own pacifist and leftist opposition movements and the USSR as well, and that stronger assertion of leadership from Washington can overcome the problem.

Even without European leaders' desire to avoid provocation, however, there is the old problem of difference in *objective* interests between Washington and NATO capitals: the American interest in a nuclear strategy that promises to keep combat away from U.S. shores as long as possible, and the allies' interest in ensuring that escalation occur as soon as possible. The Europeans' emphasis on the political component of strategy over the operational one goes logically with the emphasis on rhetoric about extended deterrence and the import of declaratory policy as a pillar of such deterrence. This long-standing latent tension in the alliance explains some of the confusion over whether the greater "visibility" of GLCMs (as opposed to SLCMs or old SLBMs assigned to Supreme Allied Commander, Europe, or SACEUR) was of real concern to European leaders or simply anticipated by U.S. planners. The domestic political constraints faced by European governments in the wake of the 1979 LRTNF decision, in fact, made visibility more of a liability than an advantage.[63]

Domestic politics have militated against serious theater nuclear arms control negotiations on both sides of the Atlantic. The potential for wringing a reasonable agreement out of Moscow depends to some extent on the durability of NATO deployment plans for the GLCM and the Pershing II, yet the advent of negotiations could give the NATO "swing" countries an excuse to bail out of participation in the deployment. (General Bernard Rogers, Supreme Allied Commander in Europe, has voiced this concern.[64]) In the United States, Carter wanted to postpone

---

Print, Staff Report to the Subcommittee on European Affairs of the Senate Foreign Relations Committee, 96 Cong. 1 sess. (GPO, 1979), pp. 8–10.

63. On NATO's decisionmaking process for the 1979 LRTNF modernization decision, see Gregory F. Treverton, "NATO Alliance Politics," in Betts, ed., *Cruise Missiles*, pp. 415–41.

64. Michael Getler, "Cruise Missiles on Subs Will Hamper NATO Plan, Rogers Says," *Washington Post*, October 6, 1981.

theater negotiations until after 1980 election; not only was he pressed by Reagan's campaign criticisms of the softness of U.S. defense policy, but there was also the worry that early negotiations would tip the Belgian decision against deployment.[65] And during the current administration meaningful compromises seem unlikely.

The complications within the alliance simply increase the difficulties in negotiating with the Russians. Earlier fears among arms controllers that cruise missile technology would lead to horizontal proliferation of modern nuclear delivery systems appear now to be unfounded. Britain and France backed away from cruise missiles (after some initial interest) as a solution to their nuclear modernization requirements,[66] alleviating Soviet concern about U.S. technology transfers that might circumvent SALT. Fear that third world countries will seek sophisticated cruise missiles, on the other hand, does not appear economically or militarily logical. For third world countries with limited resources, smaller and less diversified force structures, and contiguous enemies, multipurpose tactical aircraft with greater payload, flexibility, and reusability are a far more rational investment. Even if the propulsion and guidance technologies of the cruise missile itself do not constitute a barrier, TERCOM mapping requirements and computer support for mission planning systems are daunting.

For the Soviet Union, the prime issue now is not horizontal proliferation among many of its enemies, but vertical proliferation of American cruise missiles in Europe or at sea. The new theater arms control talks—or as the Reagan administration has dubbed them, intermediate-range nuclear force (INF) talks—will have to wrestle with a large number of fractious issues. One is whether to restrict the negotiations to missiles. The USSR prefers to include American forward-based nuclear-capable aircraft. Its insistence on including these forces in SALT I prevented any progress until it grudgingly agreed to table the issue. It is likely to be even more adamant in a negotiation focused on the European theater. But since the logical American response to a demand that U.S. forward-based systems be included is to demand that all comparable Soviet aircraft be included too, and because negotiating over numerous classes of tactical aircraft with ambiguous functions could prove formidably

65. John M. Goshko, "Brezhnev Hits NATO Delay on Missile Talks," *Washington Post,* August 27, 1980.

66. See Lawrence D. Freedman, "The European Nuclear Powers: Britain and France," in Betts, ed., *Cruise Missiles,* pp. 443–79.

complex, it is not inconceivable that Moscow might be persuaded that there is a mutual interest in focusing on missiles.

Even if this is accomplished Washington will probably have to consider significant compromises from its initial negotiating position if an agreement is to become possible. On November 18, 1981, President Reagan proposed canceling GLCM and Pershing II deployments in exchange for the dismantling of *all* Soviet intermediate-range missiles (SS-4s, -5s, and -20s) capable of striking Western Europe (the so-called zero option). Not only would this reduce the military advantages the USSR has been accruing since it began deploying the SS-20, but it would eliminate capabilities that nation has had for over twenty years. Moreover, Reagan's proposal would leave the Soviets no land-based missiles to counter either British and French SLBMs and IRBMs (which would be covered by neither START nor an INF agreement, while comparable Soviet systems would be) or U.S. nuclear-armed SLCMs.

In a general sense, then, the United States has two choices. It can persist with ambitious negotiating demands that will be unacceptable, on the assumption that going ahead with planned deployments will leave the West better off despite the lack of negotiated restraints on Soviet forces. Or it can work for a compromise agreement. The former choice is problematic in two respects. First, domestic opposition to LRTNF modernization in NATO countries could still prevent fulfillment of the 1979 deployment decision, in which case the improvement of theater force posture would come only from the limited number of prospective TLAM-Ns. (The availability of this option, in turn, could offer beleaguered European governments an escape hatch allowing them to mollify domestic opponents without totally abandoning LRTNF improvements.) Second, without negotiated restraints Moscow could choose to deploy much larger numbers of SS-20s or even more capable follow-on systems.

A compromise therefore appears preferable. The number of possibilities is endless. Cruise missiles, though, are likely to figure significantly in the compromise. Some GLCM deployments will remain desirable for two reasons. First, since the Pershing II is becoming far more expensive than originally estimated and since it is more threatening to Moscow (its flight time is about fifteen minutes and its warning time only four to eight minutes), the 108 missiles in that category might well be the first group of weapons to be bargained away if Washington managed to secure more important concessions than it could by trading away a comparable number of cruise missiles. Second, given the operational limitations of SLCMs discussed earlier, GLCMs should retain a military appeal to the West.

A very rough illustrative example of the contours of a compromise deal would be an agreement reducing the number of Soviet missiles and reducing the number of U.S. GLCMs deployed below the planned number of 464. The fact that the GLCM has a single warhead and no planned reload capability while the SS-20 has three MIRVs and probably at least one reload might be used as a rationale for excluding French and British forces from the agreement. Many more details and asymmetries would have to be dealt with, and the declared positions of Washington and Moscow make it unlikely that such a trade—or any agreement at all—will be achieved. But the United States will at least have to give the appearance of willingness to consider such compromises if it is not to be blamed (by Western Europeans) for the failure of negotiations—and that is necessary if hope for carrying out the 1979 deployment plan is to be kept alive.

The obstacles to any successful result in arms control have become almost insurmountable, so any additional complications could provide the coup de grâce. Deployment of large numbers of conventionally armed cruise missiles is an example, since they pose difficult problems for verification. If the Soviets deployed their own conventional cruise missiles, the situation would be even worse. While agreement on excluding such missiles from negotiation might seem feasible on grounds that they mutually cancel each other out, making the verification problems more acceptable, the West would have more reason than the East to fear cheating—clandestine substitution of nuclear warheads—because of the asymmetry in available information and the probability of exposure in Western societies. Moreover, even if Soviet counterdeployment of cruise missiles washed out the verification problem, it would also neutralize the enhanced conventional Western military capability that warrants significant expenditures on such weapons.

If NATO deploys conventional cruise missiles and the USSR does not, and if FRODs or EODs prove unpersuasive to Moscow, how can theater nuclear arms control be salvaged? One heretical possibility would be to have such conventional missiles deployed by *German* military forces alone, with SACEUR dividing the tactical labor of the alliance by allocating to the Bundeswehr or Luftwaffe the targets appropriate for those weapons. At first glance this seems to offer the worst of both worlds, given nervous Russian jokes about how the nuclear "GLCM" stands for "German-launched cruise missile." But the fact remains that West Germany, unlike the United States, Britain, and France, does not have nuclear ordnance. It should be credible to Moscow that West Germany's allies have objective reasons not to want to change that situation and that, if Bonn were to decide to develop and deploy its own nuclear weapons, the availability of cruise missile delivery systems (as opposed to others already available) would be a negligible factor.

If the Russians, the allies, or the American military object (the Air Force's lack of enthusiasm for cruise missiles might change if a sister service had the weapons while it did not), there are other possibilities, though they might be too ambitious to negotiate. Washington might simply plan and declare that certain models of NATO dual-capable aircraft were no longer assigned nuclear missions, stating that it would be logical to deploy nuclear GLCMs to release dual-capable aircraft for strikes with conventional munitions.[67] Soviet concern about verifying the nonnuclear status of conventional cruise missiles could be parried by citing observable differences in security precautions for the nuclear GLCMs. This would reinforce the basis for simplifying the negotiations by restricting them to missiles, deflecting Soviet insistence on dealing with more broadly defined forward-based systems, thus increasing the practical chances that the negotiating mechanism could cope with the issues.

A final alternative would be to make nuclear GLCM launchers dual-capable. Although this could entail a penalty in negotiating a theater nuclear balance if all the launchers were counted as nuclear, it might actually offer a deterrent combined with a flexible response capability if deterrence failed. Dual-capable GLCMs might symbolically reinforce the image of the "seamless web" of flexible response strategy. And should deterrence fail, the theater nuclear balance per se would not be as crucial as the mix of options available for escalation control. The conventional weapons would offer a discrete though limited capability for strikes on fixed targets in Eastern Europe, and the limitation would be less stringent because of the complementary tactical systems on which NATO already relies. The evolving discussion about the modern-

67. Such a sweeping change could be objectionable to nonnuclear allies who now have a stake in two-key dual-capable aircraft; the arrangement gives them participation in the nuclear deterrent—a finger on the safety catch as well as on the trigger—without responsibility. Plans call for nuclear GLCMs to be completely under U.S. control.

ization of theater nuclear forces may make the notion of dual-capable GLCMs sound bizarre. But is it really any more so than the current reliance on dual-capable aircraft? Such aircraft are much less desirable since they cannot shift quickly back and forth between conventional strike and nuclear deterrent missions; and the longer they are withheld for the latter, the more likely they are to be destroyed by Soviet conventional attacks.

Benefits, Costs, and Trade-offs

The preceding sections summarized the principal issues in each of the general areas of technology, strategy, and politics. It is evident that uncertainties and choices in these categories are to a large extent interdependent. Cruise missiles will enhance U.S. defense capabilities and they will complicate political relationships, which could in turn rebound against capabilities. To understand which variants of the missile will gain more for overall U.S. security than they will lose, more analysis is necessary. The issues are too complex to be resolved in this study, but this concluding section will highlight some of the critical ambiguities and alternatives. One place to begin is with consideration of the missiles' cost advantage relative to other choices for investment in weaponry.

Cruise missiles became important because they seemed cheap. Otherwise, reliance on ballistic missiles and aircraft would have remained the bedrock of long-range firepower. Thus the larger the prospective investment in cruise missiles, the more their true costs need examination. Reliable cost-effectiveness analysis, however, has always been difficult. On this issue it is far more complex, given the profusion of missile types, potential missions, and conceivable launch platforms, and the lack of lessons from relevant battlefield experience (the V-1 is hardly comparable).

Performance versus Price

Many variants of the modern cruise missile have been developed in the past few years because it has been easy to capitalize on the basic common elements of design—airframe, propulsion, and guidance. This has increased the apparent military versatility of the weapon, making it useful for a broad spectrum of missions. Over time, however, commonality accompanies versatility only within limits.

One limit is the relative emphasis on the two factors. The more versatility is stressed as a requirement—expanding the number of missions and performance standards for each one—the more strain is placed on commonality. It is cheaper to produce a large number of a single version of a cruise missile than to produce smaller numbers of many versions with more sophistication. Optimizing capabilities for each of many missions makes changes in design and components more attractive. The designs of the basic models now programmed are constrained by their origins, for reasons that are anachronistic. The airframe of the Tomahawk had to be designed to fit the dimensions of submarine torpedo tubes, and the trapezoidal Boeing ALCM was derived from the earlier subsonic cruise armed decoy, which had been shaped to fit SRAM rotary launcher racks. Such constraints can jeopardize performance, and they become progressively less acceptable as the original reasons for them become outmoded and as more ambitious applications are envisioned for the weapons. The Tomahawk's tubular shape is far from aerodynamically ideal. It limits the speed of response to sharp changes in altimeter readings, requiring the missile to fly higher when over jagged terrain, which increases its detectability by radar. Yet today the torpedo-tube launch mode originally planned may be the least important for the future.

According to an Air Force officer quoted by one reporter, commonality "is sometimes held out as an absolute virtue. But how much is artificial commonality, and how much is to be given up for it? The F-111 is a classic case of pursuing commonality for its own sake. We still carry 300 lb. of angle iron in the tail to accommodate carrier landings."[68] Commonality is a "technical push" criterion for cruise missile programs. If "strategic pull" criteria come to dominate—which they should if cruise missiles do become major elements of force structures—there will be greater pressure to slough off old constraints and to reconfigure future models to accord more closely with specific operational requirements. In that case cost advantages may recede along with commonality, and low cost and high military utility may diverge. One can only guess whether the divergence would be significant, but there is, at least potentially, a vicious circle that would cast doubt on the proper extent of investment in cruise missiles: cheapness drove commitment to the systems, but increased commitment ultimately vitiates cheapness.

Even if the divergence between cost and utility is not great it can be

68. Quoted in David R. Griffiths, "Proposal Set on Air-to-Surface Missiles," *Aviation Week and Space Technology*, December 29, 1980, p. 24.

significant as long as doubt persists about the cost-effectiveness of the missile in comparison with other systems, such as tactical aircraft, which are more expensive but are reusable and more flexible. In systems analysis it is difficult to incorporate the value of unit flexibility. If uncertainties about respective capabilities remain high while apparent cost differentials decline, there will be stronger reasons to favor the more flexible alternative.

The vicious circle cannot necessarily be broken simply by reducing the emphasis on versatility. Much of the attractiveness of cruise missiles comes from their capacity to substitute for other more expensive systems in performing military missions. If simplicity and commonality are emphasized more than adaptability to multiple missions, cruise missiles become less substitutable for other weapons, except at the price of reducing effectiveness. For example, the Air Force and the Navy resisted pressure (based on cost considerations) from the Office of the Secretary of Defense that they accept the Tomahawk MRASM, preferring a supersonic ramjet missile—as a complement if not as an alternative—for widening the envelope Soviet tactical air defenses would have to cover (by striking from high as well as low altitudes).[69] This echoes the argument presented by Foster and Bennett that the ALCM is more valuable in tandem with a penetrating strategic bomber than alone, because the combination places greater demands on Soviet strategic air defense investments. Thus the appeal of cruise missiles arising from their synergy with other systems reduces their attractiveness in terms of substitutability.

What Are the True Costs?

Cruise missiles do not seem so cheap when their unit cost off the production line rises and when the associated costs—especially for delivery systems—are considered. The significance of associated costs

69. "Pentagon Split on New Medium-Range Missile," *Flight International*, September 20, 1980, p. 1172. The preference of the Air Force for advanced conventional standoff missile options other than the MRASM, which was directed by Under Secretary of Defense William Perry toward the end of the Carter administration, led to a suggestion for a compromise—procurement of 200–300 MRASMs for use against hardened command, control, and communications targets, with a "follow-up competition between General Dynamics and McDonnell Douglas to develop an anti-runway weapon." Mark Hewish, "Tactical-Missile Survey Part 1: Ground Targets," *International Defense Review*, no. 6 (1980), p. 858. Such a compromise could reduce the ability to capitalize on the MRASM production "learning curve."

depends both on the scale of cruise missile deployments and on the life cycles of the delivery systems. It would be illusory to expect infinite economies of scale from the missile production learning curve, given the limited number of adaptable launch platforms. Although large-scale deployment of the ALCM reduces its unit costs, major increases in numbers beyond those planned would require dedicated platforms—that is, aircraft purchased solely as cruise missile launchers. Such new delivery systems (as opposed to B-52s, for which only modification costs are attributable because they are already paid for) would dramatically raise the true cost of each cruise missile.[70] Whether or not missile deployment can capitalize sufficiently on the sunk costs of adaptable platforms depends on strategic concepts and mission requirements and, most important, on changes in the role of the other legs of the triad.

The current surge in U.S. defense investments raises questions across the board about what changes will evolve in doctrine and force structure. This compounds the difficulty in attempting to make reliable assessments of the cost-effectiveness of cruise missiles, especially conventionally armed variants. They appear potentially applicable to a wide range of military missions, but their relative utility depends on decisions about air and naval force structures at a much more general level. Enthusiasm for cruise missiles should not be allowed to drive much larger decisions about force structure. For example, it would be a mistake to let Tomahawk loadings take precedence over air defense systems on surface ships. At the same time, a long-term view of the potential of conventionally armed SLCMs could warrant significant departures in ship design and naval force structure. Integrating cruise missiles into strategy, in short, requires judicious and carefully honed planning, rather than a leap to take advantage of the superficially obvious benefits of the weapons.

GLCMs appear to be inexpensive choices for theater nuclear force modernization only when compared to SLCMs on new, dedicated submarines rather than on partially converted old SSNs. Doubt about the latter option, because of the demands on existing submarines to perform other missions, helps to validate the GLCM choice. Full GLCM system costs per missile might also be reduced if fewer TELs were procured and provision was made for multiple refiring, although apparently this option is not being considered. Deploying a conventionally armed GLCM system[71] might make multiple reloads seem even more

70. See Baker, "Program Costs and Comparisons," pp. 102–08, 112–13.

71. See Palin, "War in Europe," pp. 202–05.

necessary for cost-effectiveness. But there is a limit to how far extra rounds might spread the costs. The number of fixed targets appropriate for cruise missiles is finite, and if the missiles are effective most of those targets (with the exception of airfields) would not have to be hit again frequently. Moreover, the incremental costs for the GLCM convoys that would be required to carry and load extra rounds might not be trivial. One way to reduce costs for conventional GLCMs would be to sacrifice mobility and operate them from shelters on air bases (where they might not be much more vulnerable than aircraft) and assume that such limitations would matter only if the bases were subjected to nuclear attack—at which point conventional cruise missiles would mean little anyway.

The smaller number of systems that make up U.S. strategic forces should make trade-offs in that area somewhat easier to grasp. But trends in cost advantage and military efficacy may be disjoined. The first phase of ALCM deployment is the period in which the maximum cost advantage is obtainable from the adaptable platforms (B-52s), but this is also the period in which the contribution of the missiles to target coverage is least impressive. While the number of ALCMs deployed rises in the 1990s, the reliability of the by-then ancient B-52 carriers declines, so a larger ALCM force is not necessarily equivalent to a larger force on target. Improving the prospects requires a new carrier with greater payload and faster flyaway time.[72] And a carrier that is new, dedicated, and more capable (and thus more expensive) reduces the ALCM's cost-effectiveness—the principal original rationale for commitment to the ALCM—relative to a manned bomber.

The platform cost per ALCM space on the B-52 would rise significantly in a wide-bodied cruise missile carrier and even more in a B-1 derivative carrier.[73] (This does not mean the wide-body would be more cost-effective; the B-1 type would have substantially greater prelaunch survivability.) This suggests that the ALCM's major cost advantage obtains only during the 1980s. Over the long term, however, a new ALCM carrier or bomber will have to be procured anyway if the air-breathing leg of the triad is to be preserved. So while the cost advantage of the ALCM will decline, the weapon will not become cost-ineffective unless either new manned bombers turn out to be *more* penetrative than cruise missiles or the air-breathing leg is abandoned. In view of the

72. See Bennett and Foster, "Strategic Retaliation."

73. Baker, "Program Costs and Comparisons."

uncertainty about how effective Stealth technology will be and the enduring obstacles to devising a faultless MX basing mode, neither of those conditions seems likely to obtain.[74] Thus even when the ALCM is no longer comparatively impressive, it will still probably be a worthwhile investment.

To shrink the apparent vulnerability in the 1980s "window" period, however, there might be temptations to proliferate first-generation ALCMs to the outside limit, increasing the total investment in air-breathing strategic forces to compensate for perceived inadequacies in the other elements of the triad. But if cruise missiles are to be important over the longer term, there is a counterincentive to restrain commitment to first-generation models and to wait and capitalize on a more capable second generation that incorporates greater fuel efficiency, range, speed, penetration aids, or other advances. But technical improvements in altitude and radar cross section may already be near the point of diminishing returns; at the point where increases in penetrativity depend on greater speed, rocket-powered systems will become more attractive.[75]

As a multipurpose family of weapons, cruise missiles do not lend themselves to evaluation by clear standards or binary choices. Any variant can be compared with other weapons within one of the relevant dimensions of analysis—technical, economic, strategic, tactical, political, diplomatic—but the large number of cruise missile variants and the interdependence of the different variables in the analysis preclude high confidence in a comprehensive assessment. The rest of this concluding section provides some fuzzy bottom lines, summarizing the conflicting trends and implications, the principal strategic choices, and the principal variables to which decisions on cruise missiles should be sensitive.

Doctrinal Adaptations

The most direct way in which to summarize the promise of the cruise missile is to distill its military advantages. These advantages may be

74. Desmond Ball presents rough computations suggesting that in the future, with both a survivable American MX system and a Soviet look-down–shoot-down air defense that would reduce cruise missile penetrativity, the costs per warhead on target favor the ICBM with MIRVs over the ALCM. He notes, though, that the value of preserving the triad would mitigate the negative inference about long-term investment in cruise missiles. "The Costs of the Cruise Missile," *Survival,* vol. 20 (November–December 1978), pp. 245–46.

75. Toomay, "Technical Characteristics," p. 52.

reduced, neutralized, or turned into liabilities when other considerations are applied, but the missile is above all a weapon.

For strategic nuclear forces cruise missiles offer less than early enthusiasts hoped or opponents concerned about arms control feared. But they offer more than can be inferred from narrowly focusing on the weapon in isolation from other elements of force structure or from those that could be feasible alternatives for maintaining strategic capabilities. ALCMs support the minimum requirement for strategic forces—assured destruction retaliatory capability—by buttressing one weakening leg of the triad (aging bombers whose penetrativity will decline) while deficiencies in a leg that is widely believed to be even weaker (fixed ICBMs) are remedied over time. Though ALCMs in the numbers planned may, by themselves, offer inadequate countervalue target coverage in the worst case (a Soviet surprise attack against bombers on day-to-day alert), they would add to it significantly in conjunction with SLBMs. Strictly in terms of assured destruction criteria, after all, the only rationale for a triad is to compensate for what many believe has happened with Minuteman vulnerability—emergence of a decisive preemptive threat against one leg. At the same time, if the counting rules arrived at in SALT II are a guide, ALCMs do not threaten arms control solutions for stabilizing deterrence and force levels. This, however, suggests the limits of the missile's contribution to military effectiveness.

The doctrinal emphasis in Presidential Directive 59, in contrast to simple assured destruction criteria, poses more ambitious demands than can be met by current capabilities, and to these the ALCM's contribution is limited. It is less troubling in bureaucraticpolitical terms than it was through most of the 1970s because the Carter administration's move to countervailing strategy, now reinforced by Reagan administration commitments to a generally strengthened force posture, has weakened resistance to procurement of a new penetrating bomber. In strategic terms, the ALCM is almost neutral. Its flexibility is not the sort stipulated in the directive because the missile depends on a delivery platform and mission planning support structure of dubious endurance. The other notable aspect of Presidential Directive 59 is the expansion of the target base (a menu of 40,000 installations in SIOP 5D, according to one report).[76] Since ALCMs replace other weapons, they offer little additional coverage, and the newly emphasized categories—command struc-

76. Desmond Ball, "Counterforce Targeting: How New? How Viable?" *Arms Control Today*, vol. 11 (February 1981), p. 2.

ture targets—would probably be spared beyond the firing time for ALCMs. However, in light of the overall inadequacy of current U.S. capabilities to implement the directive's requirements for endurance and flexibility, the ALCM's limitations for those purposes do not discredit the weapon. And if its accuracy can be reasonably high, the ALCM offers an incremental improvement in the second-strike counterforce requirements of the countervailing strategy during the period between the erosion of B-52 penetrativity and the deployment of more survivable ICBMs.

Finally, the implications for ALCMs for deterrence stability are positive on balance. They coincide with emphasis on counterforce-matching for deterrence by denial, while posing less danger of counterproductive effects (by increasing the "use 'em or lose 'em" reciprocal fear of surprise attack) than improved ICBMs would because of both their slowness and their limited potential for counterforce target coverage. And by the time more impressive counterforce ballistic missiles (MX and D-5) enter the strategic force, ALCMs should not figure significantly in the debate over stability since their contribution to the threat against Soviet forces would be secondary, if not incidental. (If the superpowers moved to deploy ABM systems, though, air-breathing forces could become more important since they would no longer appear less capable of penetration.)

For LRTNF, the GLCM represents a clear qualitative improvement in capability by virtue of its range, which allows wider coverage of targets (more than for the Pershing II), including ones within the Soviet homeland. Thus constituting a greater threat to the USSR (and as long as the Russians consider it as invulnerable to preemption as NATO planners do), the GLCM appears to offer enhanced deterrence when considered in isolation from the responses it may provoke. But it would be unrealistic to assume that GLCMs will really be invulnerable. Moreover, the political and diplomatic salience of the LRTNF modernization issue complicates the military rationales for the system. Much more than the ALCM, the GLCM aggravates the tension between unilaterally improved security and negotiated stabilization through arms control.

The change in capability that GLCMs present may be seen as a logical adaptation of NATO's deterrence to the combined impact of parity at the higher level of central strategic forces and continuing Western disadvantage at the lower level of conventional forces for war in Europe. The value of the adaptation depends to some extent on its ambiguity, since Americans may see it as a decoupling protection against escalation to intercontinental exchanges while Europeans prefer to see it as a recoupling link (in a sense validating the logic in the Soviet charge that the LRTNF modernization plan circumvents the balance established under SALT II). The adaptation, however, is more inadvertent than premeditated, since it was the availability of the weapon that prompted the decision to deploy it, not a preexisting doctrinal requirement that prompted its development.

These uncertainties make it more likely that arms control pressure to limit the GLCM's role will be persuasive. Greater commitment in Western Europe than in Washington to salvaging détente combined with adamant Soviet objections to the novel threat posed by the GLCM suggest that the full military potential of the system will not be realized. Thus it appears, not surprisingly, that the cruise missile with limited utility for increasing U.S. striking power (the ALCM) is politically unproblematic, while the one with greater military significance (the GLCM) will face more unyielding constraints. Even in the unlikely event that vigorous assertion of U.S. leadership and adept alliance diplomacy overcame the political hesitancy of NATO governments, the consequences of Soviet reactions for the balance of capabilities might constitute valid *military* grounds for accepting limitations on the nuclear GLCM. If it became certain that East-West arms control was dead rather than in doubt, however, the overall problems would be worse but the strategic calculations would be less complicated. Without arms control, it would also be harder to contain arguments for proliferating nuclear SLCMs.

The contribution of conventionally armed cruise missiles to Western defense is less certain than that of the nuclear variants. For at least a limited number of missions they offer promise, but expectations must be tempered by several questions about ultimate cost-effectiveness. First, can new conventional munitions and technical improvements in capability be developed within narrow limits of acceptable unit price? Second, can deployment modes be made compatible with arms control, so as to reduce the odds of Soviet counterdeployments that would neutralize the U.S. investment? And third, can integrated operational plans be adjusted for optimal employment of complementary weapon systems? The first depends on ingenuity in research and development and efficiency in production; the second on percipient judgment of the elasticities in Soviet security concerns, military investment, and negotiating positions; and the third on confidence in the reliability of sensitive components of the cruise missile under conditions of wartime friction when more flexible systems would allow greater margin for error. Commitments to conventional cruise missiles should depend on answering these questions in sequence.

Expanding conventional SLCMs would highlight two issues of strategy and doctrine that should be settled before deploying cruise missiles rather than letting strategy emerge by default from deployment decisions. One is the land-attack mission of the Navy. Should naval surface force structure continue to be largely determined by the requirement to mount attacks with conventional ordnance against targets on the periphery of the USSR? If not, the rationale for major investments in cruise missile launch platforms such as battleships, even for missions in the third world, is weak, since the unavailability of aircraft carriers in remote areas assumes that they are stationed in positions for war with the Soviet Union. Moreover, even under the third world intervention rationale, the logic implicit in big investments in capital ships with cruise missiles is questionable. Large expenditures of that sort, if unmatched by commensurate expansion of ground forces earmarked for such contingencies (if carriers need to be withheld to cover NATO or Korea, large numbers of ground forces might need to be too), emphasize offshore assets as sufficient for operational success. Such faith in independent naval firepower may be similar to the overconfidence in the efficacy of conventional air power of Air Force spokesmen in some past wars. And if the ship is meant to provide close support for ground troops, guns rather than cruise missiles (more appropriate for deep interdiction) are most relevant.

Second, should the United States continue to build the Navy around a few large ships with high unit capability (many eggs in few baskets) or shift to a larger number of smaller vessels? A shift might argue for keeping the battleships in mothballs (especially if the cost of their refurbishment rises much above optimistic early estimates) and parceling out SLCMs to destroyers or other small platforms that can be deployed in large numbers. And de-emphasizing large attack carriers could raise the importance of SLCMs for offensive striking power.

All of these doctrinal questions fade in significance (as constraints on force structure decisions) to the extent that defense budget limitations are relaxed. When expenditures involve only additional capabilities rather than trade-offs within existing aggregates, it is easier (though no more rational) to avoid strategic choices and indulge the simpler impulse that more—of anything—is better. Long-term sustained real growth in defense expenditures, however, cannot be assumed, even if the presidency and Congress remain in conservative hands for years. As the second half of the Reagan administration draws near, the consensus that supported a major expansion of military forces is diminishing. Indeed, some observers attribute the large increases in the 1982 defense budget to the administration's fear that the defense fervor may recede, making it vital to boost total obligational authority for investment as high as possible immediately.

If cruise missiles were independent weapons, their unit cost could make it reasonable to pay less attention to doctrinal questions, assuming that future decisions on their proportional role in military plans would come at a low premium. However, because many variants of the missile depend on expensive launch platforms—especially if proliferated beyond the numbers currently planned—their role has larger and more expensive implications for force structure. There is some advantage in avoiding rigid subordination of force structure to concepts of warfare that could prove erroneous, and letting technological serendipity and the principle of diversity in weapon systems shape doctrine. But that advantage has limits. Where the potential utility of cruise missiles is least certain, the technical tail must not be allowed to wag the doctrinal dog.

Effects on the Military Balance

In policy debate and decision, military and political considerations often converge around the concept of the military "balance." How do the forces of the United States and the Soviet Union, or of NATO and the Warsaw Pact, stack up against each other? How will new deployments affect the assessment of which side is ahead? Since the cruise missile as a generic system will affect the capabilities of virtually all elements of military power except ground forces, it will have widespread effects on the balance. What those effects will be are subject to three principal uncertainties: unilateral U.S. decisions on deployments, which will vary with determinations both of absolute cost and of opportunity costs in competing categories of procurement; Soviet efforts to counter additional U.S. deployments with additional Soviet weapons; and the possibility of arms control agreements that could temper both sides' unilateral options.

The ALCM is the cruise missile whose implications are the most predictable. Its scale of deployment is firmer than that of other variants, its cost-effectiveness is least ambiguous, and it is least likely to dramatically increase U.S. capabilities. This combination means that the ALCM program is less likely than others to be changed by U.S. planners, to provoke major Soviet responses in offensive capabilities, and to be bartered away in arms control negotiations. Under present plans ALCM deployment will maintain the efficacy of the bomber force by increasing penetrativity but will not notably increase the number of deliverable warheads. The number of warheads on ALCMs could be increased if they were deployed on wide-bodied carriers, which could lift far greater numbers than B-52s or B-1Bs could, but this option—never popular with the Air Force—has become even less favored in recent years. The counting rule for ALCMs determined in SALT II suggests that they will pose a much less thorny negotiating problem than GLCMs or SLCMs.

The GLCM's prospective effect on the balance is less certain. If the planned deployment is carried out without provoking additional Soviet counterdeployments, it would markedly improve the relative position of the United States in the balance of theater nuclear forces. Precisely because of that, however, the GLCM is more likely than any other version of the cruise missile to prompt a Soviet military reaction. Leonid Brezhnev pointedly announced on March 16, 1982, that if the planned NATO deployment were to be implemented, "a different strategic situation would arise in the world. There would arise a real additional threat to our country and its allies from the United States. This would compel us to take retaliatory steps that would put the other side, including the United States itself, its own territory in an analogous position."[77] This statement ignited fear that Brezhnev was tacitly threatening something as dangerous as putting SS-20s in Cuba. Whatever was meant, it was clearly a threat to discourage prospective U.S. improvements in LRTNF. There is also the possibility, slim though it is, that theater arms control negotiations could succeed. If so, the Reagan administration's zero option would have to be toned down and U.S. LRTNF deployments reduced in exchange for reductions in Soviet intermediate-range ballistic missiles. In this case, as suggested earlier, some number of GLCMs would probably be the last of the deployments to be traded away. Even

77. Foreign Broadcast Information Service, "Daily Report: Soviet Union," vol. 3 (March 16, 1982), p. R8.

less certain is the future impact of the nuclear SLCM. The TLAM-N could improve strategic or theater nuclear capabilities but only marginally; at the same time it would constitute a vexing problem for arms control should Moscow insist on verification standards for U.S. forces as stringent as those the Reagan administration demands for Soviet forces. Every reloadable torpedo tube might be cited as the equivalent of a launcher even though actual U.S. capability (and this assumes that official plans for the number of TLAM-Ns to be deployed in that mode are carried out) would be far less. If the TLAM-N stood in the way of an agreement constraining Soviet force expansion but added little to real U.S. striking power, it might worsen the relative American position in the balance rather than improve it.

The effect of conventionally armed cruise missiles on the balance of military power is the most uncertain. By providing greater flexibility in modes of delivery at long ranges and allowing optimal use of manned aircraft, conventional cruise missiles obviously offer a desirable expansion of tactical options. But this does not take cost-effectiveness into account, and cost-effectiveness is the crux. If conventional cruise missiles divert investment from other systems that could perform more missions more frequently and more effectively, their absolute benefits will not be net benefits. And if they precluded arms control—the verification problems could be even greater than those for the TLAM-N—what they gained for the conventional balance might be lost in the nuclear balance.

This is not to say that any of the programs should be abandoned or reduced because of these uncertainties. Indeed, if some of the uncertainties are resolved, it may turn out that some programs should be expanded. The uncertainties simply mean that, for the most problematic cases, clearer analysis of trade-offs is advisable before procurement and deployment decisions become irrevocable.

Focusing Future Choices

Cruise missile types can be assessed according to numerous balances of cost and benefit: military utility against financial cost; cost advantage from commonality versus military advantage from diversification; short-term versus long-term payoff from investment; military benefit versus arms control cost; missile capability versus platform limitations; and so on. Judgments about the missiles' potential are sensitive to the possibilities of combining technical improvement (to cope with evolving Soviet countermeasures) and cost containment; the availability of cheap, adaptable launch platforms; the relative importance of preserving the air-breathing leg of the strategic triad; the priority in doctrine of deep interdiction or attacks on well-defended targets; the development of credible conventions for counting missiles in arms control agreements; the balance between allies' increased confidence in military deterrence and their desire to salvage political détente; the prospective rates of progress in improving air defenses as opposed to improving missile penetrativity; and so forth. Until the unreachable point when uncertainties about technical data and procurement costs evaporate, military strategy becomes less artful and more scientific, and policy decisions by adversaries and allies become more predictable than our own, there is no grand model that can factor in all these choices and produce an answer to how many cruise missiles the United States should buy for military purposes or give up for political purposes.

This is analytically dispiriting but perversely revealing. When the cruise missile began to emerge in the 1970s many observers thought its implications were clear and predictable. Civilian enthusiasts believed it would be militarily good for everything, military skeptics feared it might gut commitments to preferred systems, and worried arms control advocates believed it would wreck strategic stability. The durability of the first of these early assumptions and the erosion of the second and third have produced a momentum that has outstripped the analytical basis for it. Cruise missile programs have rushed ahead faster than doctrinal concepts could rationalize them. In the late 1970s the political eclipse of domestic antidefense sentiment discredited anxiety about arms control negotiability to an extent unwarranted by the potential boomerang effects that indiscriminate deployments could have on the West's position in the military balance. Full recognition of the daunting complexity of technical, strategic, and political cost-effectiveness calculations should prompt foreign policy decisionmakers to pause and put cruise missile programs under the jeweler's glass. Wallowing in complexity is, of course, an analytical cop-out. A few particular conclusions about the variables governing choice warrant emphasis.

For strategic nuclear cruise missiles the critical uncertainty is time—how long the weapons will provide the benefits they promise. For the ALCM the issue is less its role through the mid-1980s than its role in a refurbished triad—or a dyad by default if ICBMs cannot be deployed satisfactorily—further in the future. The three central variables are the future of land-based ICBMs, the technological race between penetrativity and air defense, and the arms control race between the technological evolution of strategic weaponry and the ability of negotiating mechanisms to keep up.

The critical uncertainty for the nuclear GLCM is its compatibility with NATO solidarity—whether its clear enhancement of LRTNF capability can be accommodated to Soviet objections (at what price in other aspects of the balance?) and, if not, whether resultant Western European backsliding on deployment can be averted. But if allied pressure makes some major compromise unavoidable, cruise missiles may actually have more rather than less significance as a component in theater nuclear force modernization if the Pershing II component is bargained away. And if land-based LRTNF are truncated by trading away Pershings for political reasons, a U.S. administration bent on improving capabilities will have more incentive to deploy some form of nuclear SLCM. In that event the crucial problems would move from the political to the strategic realm. What sorts of naval platforms should be used for nuclear land attack—invulnerable but comparatively expensive dedicated SSNs, current adaptable but overburdened submarines, or more readily adaptable but exposed surface ships? And what if the USSR deploys more modern SLCM systems capable of striking North America as well as Europe?

Conventionally armed cruise missiles are the most challenging to thoughtful planning. How can judgment of their comparative utility be refined to ensure confidence in their cost-effectiveness compared with alternate weapon investments? How should major commitments to the TLAM-C and the TASM affect more general and significant choices about naval force structure? How can their benefits for conventional deterrence and defense be realized without rebounding against U.S. interests in the nuclear sphere (by demolishing arms control limitations that might restrain further expansion of the already worrisome Soviet threat to the West)?

If the hard foreign policy choices about cruise missiles had to be distilled into one principal problem, that problem would be the tension between conventional defense and negotiated nuclear stability. Does the "conventional push" trend in cruise missile programs offer benefits sufficiently large and certain to override the uncertain but possible costs for nuclear arms control verification? Any one of three eventualities could make this problem moot: (1) if enthusiasm for conventional cruise missiles was deflected and relatively few were deployed, the consequences for the verification problem could be finessed; (2) if either functionally related or external observable differences satisfactory to Moscow could be appended to conventional cruise missiles, the verification problem would be solved; or (3) if arms control died, verification of U.S. capabilities would cease to worry Washington, although monitoring of Soviet forces and preservation of the allies' contribution to the NATO deterrent would become more of a problem.

The last possibility is the biggest issue. It would be ironic if conventional cruise missiles emerged as a much more significant determinant of U.S. security or danger than nuclear-armed models, and if they wrecked arms control and Western alliance cohesion. Modern cruise missile programs, after all, were born primarily because of civilian fixation on the nuclear roles envisioned for them and on their enhancement of U.S. arms control initiatives. The alleged distinction in the nuclear sphere between strategies of deterrence and defense, often asserted by academic theorists who favor the logic of mutual assured destruction, has lost popularity in recent years as deterrence-by-denial theories underlying the countervailing strategy have come to the fore. Thus it would also be ironic if the alleged distinction came to seem relevant to the role of conventional cruise missiles, with the side effects of those weapons eroding confidence in the stability of the nuclear balance. So perhaps the principal issues that need to be resolved definitely before final decisions on what to do with the various cruise missiles emerging from development are the new, specific question of how financially inexpensive conventional cruise missiles will really be, and the old, general question of how strategically expensive the death of negotiated arms control could be.